# OXFORD ENGINEERING SCIENCE SERIES

# Nonsteady, One-Dimensional, Internal, Compressible Flows

## Theory and Applications

JOHN A. C. KENTFIELD

*Department of Mechanical Engineering,*
*Faculty of Engineering,*
*University of Calgary*

*New York   Oxford*
OXFORD UNIVERSITY PRESS

Oxford University Press

Oxford   New York   Toronto
Delhi   Bombay   Calcutta   Madras   Karachi
Kuala Lumpur   Singapore   Hong Kong   Tokyo
Nairobi   Dar es Salaam   Cape Town
Melbourne   Auckland

and associated companies in
Berlin   Ibadan

Published by Oxford University Press, Inc.,
200 Madison Avenue, New York, New York 10016

Oxford is a registered trademark of Oxford University Press

Library of Congress Cataloging-in-Publication Data
Kentfield, John A. C.
Nonsteady, one-dimensional, internal, compressible flows:
theory and applications by John A. C. Kentfield.
p. cm.   (The Oxford engineering science series ; 31)
Includes bibliographical references and index.
ISBN 978-0-19-507358-4
1. Unsteady flow (Fluid dynamics)
I. Title.   II. Series.
TA357.5.U57K46   1993
620.1'064—dc20   91-36530

# PREFACE

To the writer's knowledge there has not, until now, been a text available devoted exclusively to the subject of nonsteady, compressible, internal flow theory and the application of this theory to practical devices. The need for such a text has become apparent now that equipment is available commercially, the functioning of which depends upon nonsteady, compressible, flow phenomena. It is the writer's hope that the present work will be seen to fill the apparent gap in the available literature. The book has been arranged to be usable both as a text for graduate level courses and also as an introduction for readers wishing to become familiar with nonsteady flow phenomena and the practical applications of these phenomena. Most of the material in the first seven chapters is based on the content of a graduate level course on nonsteady compressible flow presented, by the writer, on several occasions in the Department of Mechanical Engineering of the University of Calgary. It incorporates, therefore, the benefits of some student feedback.

In order to fulfill the needs of graduate students, and their instructors, worked examples are included in addition to exercise problems. The first and second chapters are introductory in nature. The aim is to assist readers to adjust to what are, to those familiar only with the concepts of steady flow, unfamiliar events and circumstances without, in the process, creating a need to focus much attention on the finer details of algebraic manipulation. A generalized derivation of the classical method-of-characteristics as applied to one-dimensional, nonsteady, internal flows is included in Chapter 3. This owes much to the work of Professor D. B. Spalding, and his associates, at Imperial College, University of London. Chapter 4 is devoted to a study of the various boundary conditions necessary for handling a wide range of problems. Chapter 5 deals with methods of solution including both graphical and numerical procedures.

The remaining five chapters are devoted to specific fields of application of nonsteady flows. These include pipeline flows (Chapter 6), dynamic pressure-exchangers (Chapter 7), pulse combustors (Chapter 8) and the tuning of the exhaust induction systems of reciprocating, and Wankel type, internal combustion engines (Chapter 9). Additional nonsteady flow devices are described in Chapter 10. None of the final five chapters are offered as expert treatises on the respective subjects covered but rather as overviews supported by appropriate references to available literature. A link is made, in each of the final five chapters, to the basic material presented in the first half of the book. Where necessary additional, more specialized, theoretical material is included, in outline form, in each of the special topic, applications-orientated, chapters.

The material of Chapter 7, dealing with dynamic pressure-exchangers, and that of Chapter 8, pulse combustors, has been enlarged in scope by the addition, in each case, of a specialist bibliography. The aim of the bibliographies, which are intended to assist researchers working in these areas of specialty, is to illustrate the extent, and nature, of earlier work leading up to the present situation in which both dynamic pressure-exchangers and pulse combustors are in commercial production. The production status of the dynamic pressure-exchanger is, in large measure, the result of the persistence and hard work of Brown Boveri and Company, now Asea Brown Boveri (ABB). The ABB dynamic pressure-exchanger functions as an exhaust-gas-driven supercharger for internal combustion engines and is known as the Comprex. The work of the Lennox Company, in collaboration with the American Gas Association, has resulted in the commercial production of natural-gas fueled pulse com-bustors adapted to the domestic heating role. The Lennox units are of an air-heating kind. Another company, Hydrotherm, produces a pulse-combustor type domestic water heater known as the Hydro-pulse. Yet other firms manufacture small, portable, pulse-combustor heaters and also fogging units for insecticide dispersal, etc.

The writer is hopeful that by referring to most known applications of nonsteady flow between one set of covers this book will help broaden understanding and bring closer together workers specializing in the differing fields of nonsteady flow applications. Common threads, or links, connecting the various applications are the fundamental phenomena of nonsteady, com-pressible, internal, flows, and supporting theory, covered in detail in the first five chapters.

The writer wishes to thank, and acknowledge the assistance of, companies that contributed to the preparation of this book by providing illustrations and other materials. The text could not have been completed without the prior work of the numerous pioneers in the field and to whom it is dedicated. Special thanks are due to Thomas Lutz, E.T.H., Zürich, Switzerland and Abbot A. Putnam, Battelle Columbus Laboratories, Columbus, Ohio, who made substantial contributions to the specialized bibliographies of Chapters 7 and 8, respectively. The writer also wishes to thank Mrs. Karen Undseth for her careful, efficient, and untiring efforts in preparing the typescript and to acknowledge the patience of his wife in tolerating the inconveniences associated with his long term preoccupations with the task of authorship.

*Calgary*                                                              J. A. C. K.
*November 1991*

# CONTENTS

# LIST OF SYMBOLS

| Symbol | Nondimensional Form | Meaning |
|---|---|---|
| **Roman letters–lower case** | | |
| $a$ | $a' = a/a_{REF}$ | speed of sound |
| $a, b, c, d$ | | general constants identified within the text |
| $d$ | | pulse-combustor combustion zone diameter |
| $d_h$ | | hydraulic mean diameter of duct or passage $\equiv \dfrac{4 \times \text{cross-sectional area}}{\text{wetted perimeter}}$ |
| $e$ | | exponential |
| $f$ | $f' = \gamma f l/a_{REF}^2$ | friction force on fluid per unit mass of fluid (positive when acting on the fluid in the direction of $x$ increasing) |
| $h$ | | enthalpy |
| $h_c$ | | surface heat transfer coefficient |
| $l$ | | reference length |
| $m$ | | fluid mass |
| $m$ | $m' = m/a_{REF}$ | Riemann variable (Eq. (3.27)) |
| $\dot{m}$ | | mass flow rate |
| $n$ | $n' = n/a_{REF}$ | Riemann variable (Eq. (3.28)) |
| $n_L$ | | number of stages in labyrinth seal |
| $q$ | $q' = \gamma q l/a_{REF}^3$ | heat transfer rate into, or heat release rate within, a unit mass of fluid |
| $\dot{q}$ | $\dot{Q} = \dot{q}/C_V \dot{m}_1 T_1$ | heat transfer rate |
| $r$ | | radius |
| $s$ | $s' = s/R$ | entropy (note: $s'_{REF} = 0$) |
| $t$ | $t' = t a_{REF}/l$ | time |
| $u$ | $u' = u/a_{REF}$ | fluid-flow velocity (positive in the direction of $x$ increasing) |
| $x$ | $x' = x/l$ | distance from origin |

| Symbol | Nondimensional Form | Meaning |
|---|---|---|
| **Roman letters–upper case** | | |
| $A$ | $dA' = dA/A$ | cross-sectional area |
| $A, B, C, D$ | | coefficients |
| $C_f$ | | flow coefficient |
| $C_F$ | | pipe-flow friction factor |
| $C_i$ | | mass fraction of $i$th specie |
| $C_L$ | | discharge coefficient of labyrinth leak |
| $C_P$ | | specific heat at constant pressure |
| $C_S$ | | discharge coefficient of simple leak |
| $C_V$ | | specific heat at constant volume |
| $E, F, G, H$ | $E', F', G', H'$ | variables as defined in Eqs. (3.29), to (3.32) and in (3.56) to (3.59) |
| $E$ | | (Arrhenius) activation energy |
| $F_K$ | | Kearton and Keh's labyrinth seal leakage factor (Eq. (7.9)) |
| $F_{\text{FRIC}}$ | | force due to wall friction |
| $K$ | | (Arrhenius) reaction rate constant |
| $K_F$ | | flow area fraction |
| $K_r$ | | area ratio parameter of $r$th pipe at a multipipe junction (Eq. (4.44) |
| $L$ | | length of a duct or passage |
| M | | flow Mach number |
| $\dot{N}$ | | number of cells passing a fixed point per unit time |
| $P$ | $P' = (a')^{2\gamma/(\gamma-1)}e^{-s'}$ | absolute pressure |
| Pr | | Prandtl number |
| $R$ | | characteristic gas constant |
| $\bar{R}$ | | universal gas constant |
| Re | | Reynolds number (based on $d_h$) |
| $S$ | | loss coefficient at location indicated by subscript or between stations as indicated in the text |
| St | | Stanton number |
| $T$ | $T' = T/T_{\text{REF}} = (a')^2$ | absolute temperature |
| $U$ | | modified velocity variable (Eq. (5.21)) |
| $V$ | | volume (of fluid) |
| **Differential operators** | | |
| M | $M' = (l/a_{\text{REF}})M$ | modified Substantial Derivative (Eq. (3.25)) |

| Symbol | Nondimensional Form | Meaning |
|---|---|---|
| $N$ | $N' = (l/a_{REF})N$ | modified Substantial Derivative (Eq. (3.26)) |
| $S$ | $S' = (l/a_{REF})S$ | Substantial Derivative (Eq. (3.7)) |

**Greek letters**

| Symbol | Nondimensional Form | Meaning |
|---|---|---|
| $\beta$ | $\beta' = \beta/a_{REF}$ | inverse slope of characteristic as identified by appropriate subscript |
| $\beta$ | | pressure divider mass-flow ratio $\dot{m}_H/\dot{m}_M$ |
| $\gamma$ | | ratio of specific heats |
| $\Delta$ | | finite increment of subject variable |
| $\zeta$ | | pressure equalizer enthalpy ratio $(C_{PL}\dot{m}_L T_L^\circ)/(C_{PH}\dot{m}_H T_{Ho})$ |
| $\eta$ | | product of isentropic efficiencies of expansion and compression for pressure dividers and equalizers |
| $\eta_{COMP}$ | | isentropic compression efficiency |
| $\lambda$ | | general variable representing either Riemann variable |
| $\rho$ | | fluid density |
| $\mathscr{P}$ | | modified pressure variable (Eq. (5.22)) |
| $\sigma$ | | generalized pressure ratio |
| $\tau$ | | generalized temperature ratio |
| $\phi$ | | general subject variable |
| $\Phi$ | | area ratio of partly open end |
| $\Phi_{l,x}$ | | leakage function for simple leaks (Fig. 7.14) |
| $\chi$ | | modified entropy variable (Eq. (5.23)) |

**Subscripts**

| Symbol | Meaning |
|---|---|
| $A, B, C, D$, etc. | zones within a flow field |
| BOUNDARY | boundary condition |
| $c$ | conditions within a cell |
| CELL | applicable to flow in pressure-exchanger cells |
| ENTRY | cell entrance flow |
| EXIT | cell, or duct, exit flow |
| $f$ | fuel |
| FINAL | final condition in a cell or duct |

| Symbol | Nondimensional Form | Meaning |
|--------|---------------------|---------|
| H, L, M | | high, low, and medium pressure, respectively |
| IN | | inflow to pressure-exchanger rotor |
| INITIAL | | initial conditions in a cell or duct |
| INLET | | inlet to a duct or cell |
| KNOWN | | variable of known value |
| $l$ | | leakage flow |
| $m$ | | with reference to an $m$ or $m'$ Riemann variable |
| MAX | | maximum value |
| MIN | | minimum value |
| MIN(EFFECTIVE) | | minimum effective value |
| $n$ | | with reference to an $n$ or $n'$ Riemann variable |
| o | | stagnation pressure or temperature (usually in conjunction with a locational subscript) |
| OUT | | outflow from pressure-exchanger rotor |
| PORT(SW) | | port, single wave process |
| $r$ | | $r$th connection at a multiple junction |
| REF | | reference condition |
| $s$ | | with reference to an $s$ or $s'$ entropy variable |
| STAGNATION | | stagnation pressure or temperature |
| STATIC | | static pressure or temperature |
| THROAT | | throat of a choked inlet or outlet |
| UNKNOWN | | variable of unknown value |
| $W$ | | at a wall |
| $x$ | | initial steady-flow condition upstream of a duct enlargement or contraction; also static condition at the discharge of a simple leak |
| $y$ | | initial steady-flow condition downstream of a duct enlargement or contraction |
| $\theta$ | | local condition along a streamline in a curved duct |

| Symbol | Nondimensional Form | Meaning |
| --- | --- | --- |
| 0, 1, 2, 3, etc. | | local stations within the flow field as indicated, specifically, in the text |
| I, II | | regions to the left and right, respectively, of a flow-field discontinuity or interface |

**Superscripts**

| | | |
| --- | --- | --- |
| $'''$ | | per unit volume |
| * | | prior to influence of leakage |

**Abbreviations**

| | | |
| --- | --- | --- |
| $a, b, c$, etc. | | wave diagram regions and corresponding state values |
| BCD | | bottom, or inner, dead-center |
| C | | turbocompressor |
| $e_1, e_2, e_3$, etc. | | end-of-duct stations in a pulse combustor (Fig. 8.6) |
| EC | | exhaust valve closes |
| EO | | exhaust valve opens |
| HP, LP, MP | | high, low, and medium-pressure region, respectively |
| IC | | inlet valve closes |
| IO | | inlet valve opens |
| L.H.S. | | left-hand side (of an equation) |
| R.H.S. | | right-hand side (of an equation) |
| T | | turbine |
| TC | | transfer closes |
| TDC | | top, or outer, dead-center |
| TO | | transfer opens |
| STOICH. | | stoichiometric |

# Nonsteady, One-Dimensional, Internal, Compressible Flows

# 1

# INTRODUCTION

## 1.1 The Occurrence of Nonsteady Flows

Nonsteady flows are generated in a very large number of practical situations. For example, turbulence associated with a so-called steady flow is in fact a random, multidirectional, nonsteady flow. The formation of a Karman vortex street in the wake of a bluff body, such as a cylinder with its axis normal to the direction of the oncoming stream, is an essentially two-dimensional, nonsteady, flow phenomenon derived from a steady upstream flow. The vortex shedding frequency is correlated, for the Reynolds number range over which vortex shedding occurs, in terms of the flow velocity, and a dimension of the body, by the Strouhal number.

Yet another example of nonsteady flow occurs in turbomachinery. Here the pressure and velocity gradients, for example, prevailing in the space between adjacent blades in a rotating component, say a turbine or compressor, appear to an observer moving with the blading, to be representative of a steady flow provided any turbulence is ignored. To a stationary observer the same situation appears to be of the periodic nonsteady kind; the pressure and velocity both vary cyclically as successive blades pass the observer. Conversely, the flow through the stator appears to be nonsteady to an observer moving with the rotor.

A fairly common, noticeable, and unwanted, manifestation of what is essentially one-dimensional nonsteady flow sometimes occurs in domestic water systems and elsewhere. It is identified by the descriptive label of water hammer. This phenomenon is associated with the longitudinal transmittal, and subsequent reflection, of pressure waves along the water column contained in the pipe system. Water hammer is usually initiated by the sudden opening, or more commonly the sudden closure, of a valve.

One-dimensional nonsteady flows of compressible fluids, the topic upon which attention is to be focused, occur in the inlet and exhaust systems of reciprocating internal combustion engines and also in devices especially designed to utilize nonsteady phenomena. Devices of the latter type include shock tubes, pressure-exchangers, pulse combustors, and nonsteady flow ejectors; the functioning of each of these will be described in detail later. Unwanted one-dimensional, non-steady flows often occur in systems such as natural-gas pipelines, etc., during transients associated with changes in operating conditions. Other unwanted one-dimensional, compressible, nonsteady flows can occur in the outflow from reciprocating, and some rotary, compressors.

Serious consequences of this situation usually take the form of destructive vibration of the delivery pipe and the transmittal of vibrations to other equipment.

The current level of knowledge in the field of one-dimensional, non-steady, compressible flows is such that it is inherently possible to make predictions, with reasonable accuracy, for many practical situations. This has resulted in a greater awareness of both the benefits to be obtained from the constructive and creative use of nonsteady flow phenomena and also the penalties that can be incurred when these are ignored. With the possible exception of the design, use, and study of wind musical instruments, most conscious attempts to utilize nonsteady flows to perform useful tasks appear to be of fairly recent origin.

## 1.2  Background

What was probably the first attempt to utilize essentially one-dimensional nonsteady flow for practical purposes was, according to Gibson (Gibson, 1954) originated by Montgolfier at the end of the eighteenth century. This apparatus was a flow-operated pump known as a hydraulic ram; a simple hydraulic ram is illustrated diagrammatically in Fig. 1.1. It is worthwhile studying, briefly, the

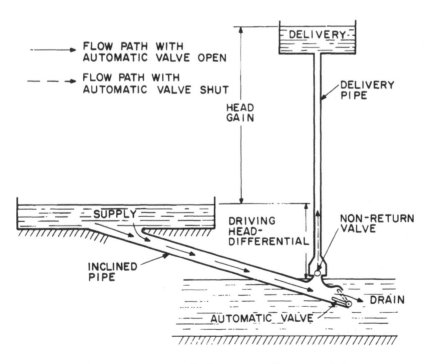

**Fig. 1–1** Hydraulic-ram nonsteady-flow pump (diagrammatic).

cycle of this device because, in many ways, features of its operation are representative of modern nonsteady, compressible, flow equipment.

In the hydraulic ram a column of water flows through an inclined pipe from an elevated source and discharges at a lower level. The water column is brought to rest by the comparatively sudden closure of an automatic valve at the downstream end of the inclined pipe. A transient overpressure effect, due to the inertia of the water column, is utilized to eject a relatively small flow of water via a nonreturn valve equipped delivery pipe. The delivery pipe discharges at a height greater than that of the supply. When the transient overpressure has dissipated the automatic valve at the downstream end of the inclined pipe reopens, due to gravity, and flow is re-established through the inclined pipe before the cycle is repeated. The essential point of interest is that, without the intervention of any moving parts other than the two valves, a sufficiently high internal pressure is generated transiently to elevate a portion of the flow to a head in excess of that of the supply. The generation of pressures greater than those of the supply is a common feature of many devices utilizing nonsteady compressible flow. In fact, there is a much more recent apparatus, known as a pressure-exchanger divider, which is the compressible flow counterpart of the (essentially) incompressible flow hydraulic ram. The pressure-exchanger divider is described in Chapter 7.

The first intentional uses of nonsteady compressible flow occurred in the early part of this century with the development of pulse-jet-like gas generators by several pioneers. Fuller descriptions of this early work have been given by Foa (1960), Edelman (Edelman, 1947), and Putnam et al. (1986).

Following the early work on pulsating combustion interest in utilizing nonsteady compressible flow seems to have waned until approximately 1930 when the development of turbochargers for diesel engines and, later, aircraft engines resulted in attention being paid to related nonsteady flow effects, particularly in exhaust manifolds, and the influence of these effects on the operation of the turbines of turbochargers. This general area of activity has subsequently developed and today represents a classical application in the non-steady, one-dimensional, compressible flow field. The range of application has also broadened to include the "tuning" of the inlet and exhaust systems of piston engines, particularly of the normally aspirated type, used for racing. The application of the principles of nonsteady compressible flow to flow within the induction and exhaust systems of reciprocating internal combustion engines receives attention in Chapter 9.

Before and during the Second World War further work was carried out, in Germany, on the pulsejet concept resulting in the development by Schmidt, and production by the Argus company, of a valved pulsejet engine used as the propulsion unit of the well-known "V1" flying bomb (Thring, 1961). Immediately followed the Second World War the French national aeroengine manufacturing company, SNECMA, developed a range of high performance valveless pulsejets. Also during this period and earlier Shultz-Grunow applied the method-of-characteristics to the analysis of pulsejet flows. This analytical tool was subsequently extended to a wider range of problems by later workers (Foa, 1960). The use of the method-of-characteristics represented an enormous advance in analytical capability because it permitted a detailed understanding to be gained of non-steady, compressible, flow phenomena. However, prior to the development of fast digital computers, and adaptation of the method-of-characteristics to numerical computation, the procedure was generally found to be too slow and tedious for day-to-day use.

During the Second World War Seippel, of the Brown Boveri Company of Switzerland, and Jendrassik of the Gantz Company of Hungary, separately conceived nonsteady flow machines utilizing cellular rotors. Devices of this type were later categorized under the umbrella label of dynamic pressure-exchangers (Chapter 7). Such machines were subsequently developed for a number of applications. The persistent work of Brown Boveri has resulted in their machine, identified by the name "Comprex," emerging as an efficient, effective, and operationally flexible exhaust-gas-energy driven supercharger for diesel engines. Comprex super-chargers are now in commercial production.

Recent emphasis in the continuing development of pulse combustors has resulted in increasing attention being paid to the development of pulse combustors for domestic and commercial heating purposes. The driving force for this activity is the ever-increasing cost of fossil fuels. Because of the relatively high rates of heat transfer achievable with nonsteady flow devices, pulse-combustor heaters have been constructed that are substantially more efficient than conventional steady-flow heaters having similar heat transfer areas. Pulse combustors, several types of which are in commercial production for heating and other applications, are considered in more detail in Chapter 8.

Before progressing from this very brief and somewhat incomplete history of nonsteady compressible flow technology to a consideration of the analytical techniques available in the area, it is worthwhile pausing to obtain, as simply as possible, an indication of the feasibility of nonsteady compressible flow processes from a fundamental viewpoint. This task can be completed, without any knowledge of wave mechanics, using only the basic concepts of the First Law of Thermodynamics and the conservation of mass.

## 1.3  Physical Considerations

Consider the filling, from a large reservoir such as the atmosphere, of the rigid, adiabatic, vessel shown in Fig. 1.2. When the initial pressure of the fluid at rest within the vessel is less than that of the surroundings the opening, and subsequent closure, of valve X will permit some fluid to enter the vessel. Applying the First Law of Thermodynamics to this filling process when the inflow terms are referred to stagnation conditions, which are the surroundings conditions, and hence the need to consider the kinetic energy of the flow passing through valve X is eliminated:

*flow work + input of internal energy = increase of internal energy*
*of contents of vessels*

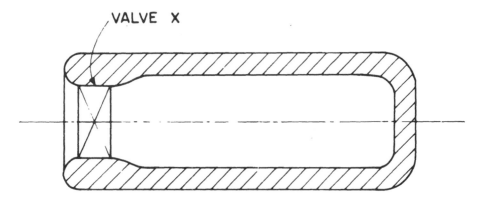

**Fig. 1-2** Rigid adiabatic vessel, or container, which can communicate with its surroundings via valve X.

i.e.,

$$P_1 V_1 + m_B c_V T_1 = m_C c_V T_2 - m_A c_V T_0 \tag{1.1}$$

where

$m_A$ = mass of fluid in vessel initially,
$m_B$ = mass of fluid added to vessel during filling,
$m_C$ = mass of fluid in vessel after filling,
subscript 0 refers to initial conditions in vessel,
subscript 1 refers to stagnation conditions of inflow,
subscript 2 refers to final conditions in vessel.

Since, $c_P - c_V = R$, and $P_1 V_1 = m_B R T_1$, substitutions in Eq. (1.1) give:

$$m_B c_P T_1 = m_C c_V T_2 - m_A c_V T_0 \tag{1.2}$$

for the special case where $m_A = 0$, conservation of mass gives:

$$m_B = m_C \tag{1.3}$$

and hence substituting from (1.3) in (1.2) and rearranging:

$$\frac{c_P}{c_V} = \frac{T_2}{T_1} \tag{1.4}$$

If the filling process is assumed to be isentropic:

$$\frac{P_2}{P_1} = \left(\frac{T_2}{T_1}\right)^{\gamma/(\gamma-1)} \tag{1.5}$$

hence from (1.4) and (1.5) after invoking the definition of $\gamma$:

$$\frac{P_2}{P_1} = \gamma^{\gamma/(\gamma-1)} \tag{1.6}$$

By way of example, when $\gamma$ has the commonly accepted value for air of 1.4:

$$\frac{P_2}{P_1} = 3.25$$

At first glance it may seem surprising that $P_2/P_1$ exceeds unity. However, this result is consistent with the observation that a hydraulic ram (Section 1.2) can generate internal pressures that transiently exceed the stagnation pressure of the supply. It is apparent that the result given by Eq. (1.6) implies precise control of the time of closure of valve X in order to trap the maximum quantity of fluid within the vessel. Further thought suggests that not only must the valve closure be timed correctly but also, to avoid throttling losses, both the valve opening and closure must be instantaneous.

Performing a similar analysis for an emptying process based on the same notation as before with the exceptions that $m_B$ represents mass removed from the vessel and subscript 1 refers to an outflow:

$$P_1 V_1 + m_B c_V T_1 = m_A c_V T_0 - m_C c_V T_2$$

Hence for the special case where $m_C = 0$ it follows that for an isentropic process:

$$\frac{P_0}{P_1} = \gamma^{\gamma/(\gamma-1)} \tag{1.7}$$

This result could have been obtained directly from Eq. (1.6) since reversible processes were postulated.

Because details of the wave mechanics were ignored in the preceding analyses all that can really be deduced is that the concept of attempting to use transient flow phenomena to establish within a vessel, such as that shown in Fig. 1.2, a final pressure greater than that of the supply or conversely, for an emptying process, a final pressure lower than that of the surroundings does not contravene the First Law of Thermodynamics. Clearly more comprehensive analysis is needed to obtain a real understanding of nonsteady, one-dimensional, compressible flows. Several analytical techniques are available.

## 1.4   Analytical Methods

The forms of solution adopted for solving one-dimensional nonsteady flow problems usually depend upon a combination of physical circumstances and the required accuracy of the solution. For cases where transients occur slowly, in relation to the time taken for an acoustic signal to travel through the system being analyzed, quasi-steady flow techniques can be applied. For other cases characterized by small perturbation amplitudes well-established classical acoustical methods are commonly used. However, a more general technique applicable to flows with large perturbation amplitudes is the method-of-characteristics; this analytical tool was referred to briefly in Section 1.2.

Attention will be focused here on the method-of-characteristics, and adaptations of it to suit numerical procedures, as the predictive technique of prime importance. However, a more elementary theory, similar in many respects to the method-of-characteristics, will be introduced first. This will allow readers who are completely unfamiliar with nonsteady, one-dimensional, compressible flow to acquire, with comparatively little effort, a familiarity with many of the relevant phenomena and to solve fairly simple problems very quickly without the need for numerical computation or tedious graphical procedures. The elementary theory is presented in Chapter 2.

The equations needed for implementing the method-of-characteristics are derived in Chapter 3 and several simple illustrative problems are solved. Chapter 4 is devoted to the derivation and manipulation of boundary conditions required for sophisticated applications of the method-of-characteristics. Solution procedures, both graphical and numerical, for problems other than those of the most basic kind are discussed in Chapter 5.

Application of the method-of-characteristics to practical problems is dealt with in later chapters. One chapter is devoted to each class of problem. In Chapter 7, which deals with pressure exchangers, a description is given of a very simple and easy to use analytical procedure that is applicable only to solving pressure-exchanger problems. This method offers much of the simplicity and speed of the elementary theory described in Chapter 2 but when compared with the elementary theory brings, to the prediction of flows in pressure exchangers, a much improved accuracy.

# 2

## ELEMENTARY THEORY

The elementary, and fairly well-known, theoretical treatment presented here is less comprehensive than the method-of-characteristics that is introduced in Chapter 3. However, elementary theory does allow the reader to gain a familiarity with many aspects of idealized, nonsteady, compressible flow behavior. The price to be paid for the simplistic approach is the imposition of limitations on the scope of problems that can be tackled. These limitations are that normally:

a. analysis is restricted to isentropic pressure waves only,
b. all waves are of infinitesimal width and therefore propagate as discontinuities,
c. flows are restricted to ducts of uniform cross-sectional area,
d. only fully open or fully closed boundary conditions can be handled at tube ends,
e. wave-on-wave interactions are not considered,
f. only a single working fluid of uniform entropy can be taken into account.

It should be clear from the foregoing that the influences of friction and heat transfer are generally excluded.

A type of flow situation representative of those that can be handled is a filling process similar to that described, in the previous chapter, in relation to Fig. 1.2. Consider, for example, the filling of a tubular container, or vessel, of uniform cross-sectional area with a valve at the end capable of opening, or closing, instantaneously and having an unobstructed flow area equal to the cross-sectional area of the container. A container of this type, which is assumed to have adiabatic walls, is shown in Fig. 2.1(a). When the pressure of the fluid at rest in the vessel is less than that of the surroundings sudden opening of valve X (Fig. 2.1(a)) at the left-hand end of the vessel results in fluid rushing into the vessel and initiating a rightward moving discontinuity or compression wave. In front (i.e., to the right) of the wave the fluid is that which was in the vessel originally and it is at rest. Behind (i.e., to the left) the wave the fluid has been compressed, and set in uniform motion to the right, while fluid from the surroundings flows into the tube through the fully open valve X. The compressive

10

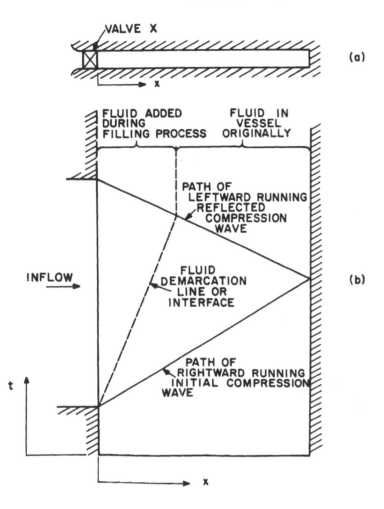

**Fig. 2–1** A typical gas dynamic filling process (a) tubular container with valve at left-hand end, (b) wave ($x \sim t$) diagram for filling process.

wave propagates rightward faster than the rightward flow velocity prevailing behind the wave. In the limit, for infinitesimally weak waves, the wave-propagation velocity will be the acoustic velocity in the undisturbed flow. When the rightward moving wave reaches the closed end of the tube all the fluid in the tube, that is the original contents of the tube plus the fluid that flows in through valve X, is traveling to the right at the inflow velocity through valve X.

Since the rightward flow obviously cannot continue through the closed right-hand end of the tube, a reflected compression wave is generated at the right-hand end wall. This serves to compress, further, the contents of the tube

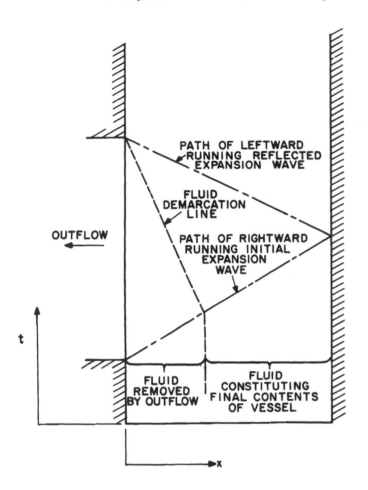

**Fig. 2-2** Wave diagram for a typical emptying process.

and to bring the flow to rest. The reflected wave propagates leftward against the direction of inflow. Provided there is no entropy discontinuity between the fluid added to the tube and the original contents the fluid interface is transparent to the reflected leftward moving wave that continues to propagate until it reaches valve X. If at this moment valve X is closed instantaneously the fluid trapped within the tube will all be at rest and at a higher pressure than that of the surroundings. The feasibility of such a phenomenon was investigated previously (Section 1.3) and it was found that it was consistent with the requirements of the First Law of Thermodynamics.

The sequence of the events during the filling process can be plotted on a distance (x), time (t) basis, otherwise known as a wave diagram, and this has been done in Fig. 2.1(b). In Fig. 2.1(b) a closed end is represented by shading and an open end by the absence of shading. Hence the period during which valve X should be open can be deduced immediately from the $x \sim t$ diagram provided the wave-propagation velocities are known.

The reverse process to filling, that is emptying where the initial pressure in the vessel exceeds that of the surroundings, is depicted on the $x \sim t$ plane in Fig. 2.2. Here the illustration of the vessel, shown in Fig. 2.1(a), has been omitted since it contributes nothing to the $x \sim t$ diagram showing the physical events. Essential differences between the filling and emptying processes are that the pressure waves occurring during emptying are both of the expansion type and, with valve X located on the left, outflow occurs to the left.

It can be seen that had the vessel of Fig. 2.1(a) been drawn with the valve at the right-hand end, and the closed end at the left, the flow and wave propagation directions would be the reverse of those shown in Figs. 2.1(a) and 2.2. The nature of the wave events occurring would otherwise be the same.

In a real device valve X cannot be expected to open or close instantaneously nor can friction between the moving fluid and the internal surface of the tube be eliminated. However, apart from these differences, the situations pictured in Figs. 2.1(b) and 2.2 can be shown to approach the conditions corresponding to limitations (a) to (e) inclusive for weak pressure waves. Restriction (f) can be satisfied by choosing, for the filling problem, a fluid constituting the initial contents of the vessel that is the same as, and of equal entropy to, the surrounding fluid.

Having obtained an impression of the physical nature of the flow for the particular cases of filling and emptying it is now possible to analyze, quantitatively, a more general situation where a single, weak, isentropic, pressure wave propagates through fluid in a region where the flow velocity in the $x$ direction is $u$ in front of the wave.

## 2.1   Derivation

Applying the conservation of mass principle between fixed stations A and B of the duct shown in Fig. 2.3 during a time interval $\delta t$ in which a weak pressure wave travels an elemental distance $\delta x$.

*Net mass inflow during $\delta t$ = mass accumulation during $\delta t$*

i.e.,

$$(u + \delta u)(\rho + \delta \rho)A\delta t - \rho u A \delta t = \delta \rho A \delta x \qquad (2.1)$$

but

$$\delta x = (a + u)\delta t \qquad (2.2)$$

thus from (2.1) and (2.2):

$$(u + \delta u)(\rho + \delta \rho) - u\rho = \delta \rho(a + u)$$

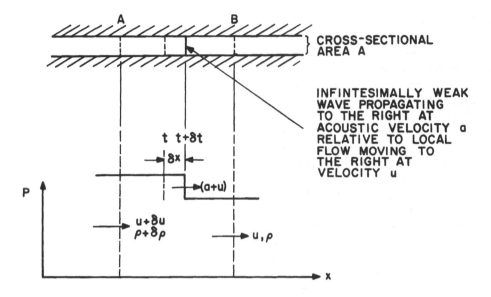

**Fig. 2–3** Elemental isentropic wave propagating to the right in gas flowing in the same direction.

or after expansion:

$$u\rho + \rho\delta u + u\delta\rho + o(\delta^2) - u\rho = a\delta\rho + u\delta\rho$$

Thus after simplification and ignoring terms of order $\delta^2$

$$\delta\rho u = a\delta\rho$$

or

$$\delta u = a\frac{\delta\rho}{\rho} \tag{2.3}$$

Assuming that the wave is isentropic:

$$\left(\frac{a}{a_{REF}}\right)^{2/(\gamma-1)} = \frac{\rho}{\rho_{REF}} \tag{2.4}$$

and from logarithmic differentiation of Eq. (2.4)

$$\left(\frac{2}{\gamma-1}\right)\frac{\delta a}{a} = \frac{\delta\rho}{\rho} \tag{2.5}$$

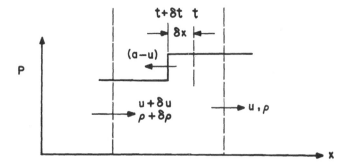

**Fig. 2–4** Elemental isentropic wave propagating to the left in gas flowing in the opposite direction.

substituting for $\delta\rho/\rho$ in Eq. (2.3) from (2.5)

$$\delta u = \left(\frac{2}{\gamma - 1}\right)\delta a \tag{2.6}$$

Proceedings to the limit where $\delta$ is vanishingly small and then integrating Eq. (2.6), without limits, and subsequently rearranging:

$$a - \left(\frac{\gamma - 1}{2}\right)u = \text{CONSTANT} \tag{2.7}$$

For a compression wave propagating against the flow, as depicted in Fig. 2.4, it can be shown by a similar analysis that

$$a + \left(\frac{\gamma - 1}{2}\right)u = \text{CONSTANT} \tag{2.8}$$

The results given by Eqs. (2.7) and (2.8) can be shown to cover all individual, or simple, isentropic waves including expansion wave cases not represented in either Figs. 2.3 or 2.4. Thus combining Eqs. (2.7) and (2.8)

$$a \pm \left(\frac{\gamma - 1}{2}\right)u = \text{CONSTANT} \tag{2.9}$$

Equation (2.9) is the required result; the positive or negative signs arise, for positive $u$, as indicated in Fig. 2.5.

Three questions can well be asked relating to the analysis. Two of these derive from the fact that only the principle of the conservation of mass and the assumption of an isentropic process were used to derive Eq. (2.9). How then can assurance be obtained that for the elemental wave:

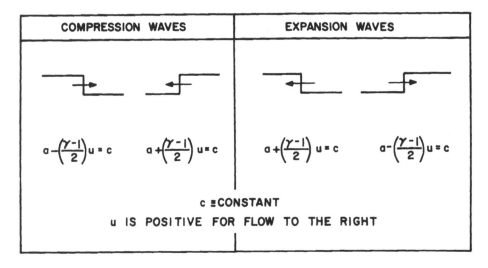

Fig. 2–5 Isentropic pressure waves and corresponding equations.

  i. momentum is conserved?

  ii. energy is conserved?

The remaining question is:

  iii. what is the meaning, in physical terms, of Eq. (2.9)?

  The answer to the first two questions is that it can be shown that the analysis is compatible with the principles of the conservation of both momentum and energy. It is suggested, in the problems at the end of the chapter, that readers verify this for themselves. An answer to the third question is that a particular value of the constant on the R.H.S. of Eq. (2.9) applies across (i.e., at the foot, crest, and at any intermediate point) an isentropic wave. Because integration was carried out the results represented by Eq. (2.9) apply to isentropic waves of any strength, or pressure ratio, not merely infinitesimally weak waves. In fact the two forms of Eq. (2.9) are collectively known as the Riemann invariants (Courant and Friedrichs, 1948) when considered from the viewpoint of the constancy of the R.H.S. for a specific wave or, alternatively, they are identified as the Riemann variables (Rudinger, 1969) when considered from the viewpoint that they are important variables in the analysis of nonsteady compressible flow fields. In each case the reference to Riemann is in recognition of his pioneering work in nonsteady, compressible, flow theory (Riemann, 1858/59).

## 2.2  Construction of $u \sim a$ Diagrams

Equation (2.9) shows that, on a $u \sim a$, or state, plane, all possible combinations of $a$ and $u$ for a particular value of the CONSTANT can be represented by

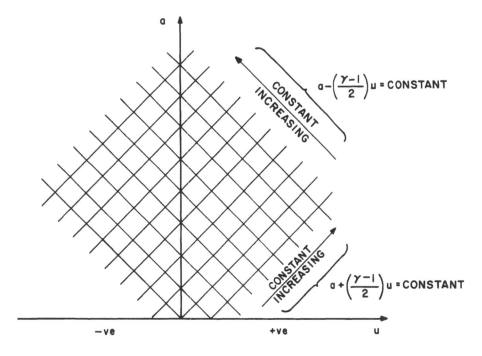

**Fig. 2–6** State $(a \sim u)$ diagram with sample change-of-state paths.

two inclined straight lines—one with a positive slope, the other with a negative slope. If the $a$ (ordinate) scale is made $2/(\gamma - 1)$ times that used for the $u$ (abscissa) scale, then one line has a slope of $+45°$, the other a slope of $-45°$. Plotting similar pairs of lines for other discrete values of the CONSTANT on the R.H.S. of Eq. (2.9) results in a mesh of lines having slopes of $\pm45°$ as shown in Fig. 2.6.

It is often convenient to convert $a$ and $u$ to dimensionless forms. This can be done by dividing both $a$ and $u$ by an arbitrary acoustic velocity, $a_{REF}$, corresponding to a convenient reference condition. Thus the variables preferred to $a$ and $u$ are

$$a' = \frac{a}{a_{REF}}, \qquad u' = \frac{u}{a_{REF}}$$

Before $u' \sim a'$ diagrams can be constructed for realistic situations it is essential to know how to represent boundary conditions on the $u' \sim a'$ plane. Because of the limitations imposed by the elementary treatment, only very simple boundary conditions need be considered. These are:

i. closed end: $u' = 0$

ii. constant static pressure condition (can apply at an open end): since all processes are assumed to be isentropic

$$\left(\frac{a}{a_{\text{REF}}}\right)^{2\gamma/(\gamma-1)} = \frac{P}{P_{\text{REF}}} = a'^{\,2\gamma/(\gamma-1)}$$

hence a constant value of $P$ is represented by a constant, or uniform, value of $a'$.

iii. constant total, or stagnation, pressure condition (can apply at an open end):
since it is normal to assume quasi-steady-flow conditions apply at boundaries, it can be shown from the steady-flow energy equation

$$c_P T_0 = c_P T = \frac{u^2}{2}$$

that

$$a_0'^{\,2} = a'^{\,2} + \left(\frac{\gamma-1}{2}\right) u'^{\,2}$$

and since

$$\frac{P_0}{P_{\text{REF}}} = a_0'^{\,2\gamma/(\gamma-1)}$$

a constant value of $P_0$ corresponds to a constant value of $a_0'$.

Usually it is normal to assume that for an outlet without a diffuser the static pressure, $P$, is constant [i.e., boundary condition (ii)] and for a loss-free inlet $P_0$ is constant [i.e., boundary condition (iii)].

## 2.3  Examples of $u' \sim a'$ Diagrams

The $u' \sim a'$ (state) and accompanying $x \sim t$ (wave) diagrams for a typical filling process, such as that illustrated in Fig. 2.1, are presented in Fig. 2.7. Figure 2.8 shows the corresponding $u' \sim a'$ and $x \sim t$ diagrams for an emptying process, such as that depicted in Fig. 2.2. It can be seen from both Figs. 2.7 and 2.8 that the labeling identifying the corners of the $u' \sim a'$ diagrams also represents the corresponding flow regimes between the waves on the $x \sim t$ plane. It is implicit, from the proportions of the $u' \sim a'$ diagrams of Figs. 2.7 and 2.8, that the $a$ scale has been made $2/(\gamma-1)$ times that used for $u'$ and also, to improve clarity, compression waves are represented as solid lines and expansion waves as chain-dotted lines on the $x \sim t$ plane.

Figure 2.9 shows a simplified model, consistent with elementary theory, of flow events in a shock tube, a device fairly commonly used, in more sophisticated forms, for physical chemistry and fluid mechanics research. For the shock tube shown in Fig. 2.9 the left-hand portion is initially maintained at a higher pressure than the right-hand section by means of a frangible diaphragm at

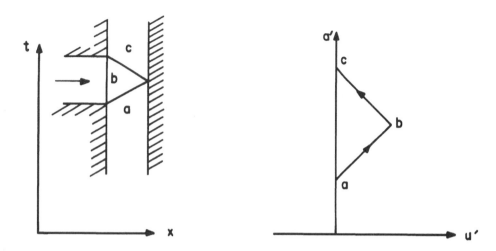

**Fig. 2–7** Filling process on $x \sim t$ and $a' \sim u'$ diagrams.

station AA. After the diaphragm has been ruptured instantaneously, a compression wave (in reality a shock wave), advances to the right and an expansion wave propagates to the left. The compression and expansion waves subsequently reflect from the closed right- and left-hand ends of the shock-tube, respectively.

An example of flow involving wave reflections at open ends is shown in Fig. 2.10; here two very large reservoirs, A and B, are connected by a pipe of length L. The pressure in A exceeds that in B. A valve, X, which can be fully opened instantaneously is located adjacent to vessel A. It is assumed that A

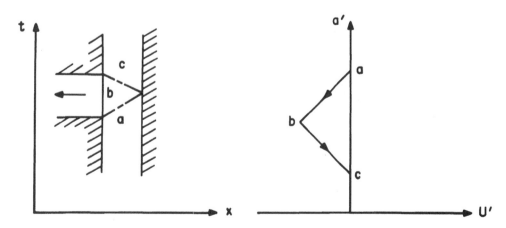

**Fig. 2–8** Emptying process on $x \sim t$ and $a' \sim u'$ diagrams.

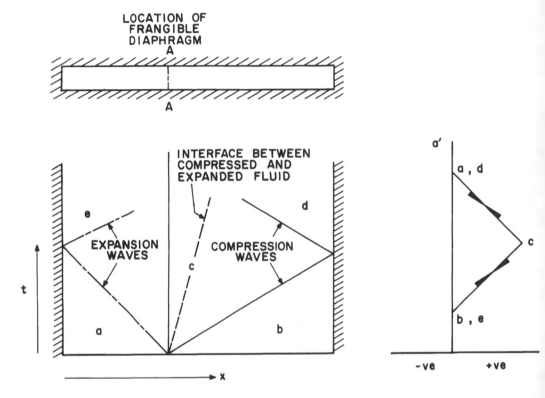

**Fig. 2–9** Wave ($x \sim t$) and state ($a' \sim u'$) diagrams for a simple shock-tube with diaphragm located at AA.

and B are each of such a size that the pressure changes that occur within them, after valve X is opened, are quite negligible over the duration of transient flow in the pipe. As can be seen from Fig. 2.10 when the valve is opened a compression wave travels from X to Y where it is reflected as an expansion wave that propagates, against the flow, back to X, where it is reflected as a compression wave, and so on until, finally, steady flow develops from reservoir A to reservoir B. It is assumed that there are no inlet losses and hence the stagnation pressure of the flow entering the pipe is equal to $P_A$. It is also assumed that there is no diffuser at station Y consequently the static pressure at Y is equal to the pressure, $P_B$, in reservoir B. The situation which prevails for the same configuration when $P_B$ exceeds $P_A$, and hence flow is from Y to X, is shown in Fig. 2.11. The essential difference between this case and that depicted in Fig. 2.10 is that when $P_B$ is greater than $P_A$ the pipe is filled initially with high pressure fluid and consequently the first wave is an expansion wave.

In each of the foregoing examples the state of the working fluid can be determined from the $u' \sim a'$ diagrams. It is for this reason that $u' \sim a'$ diagrams

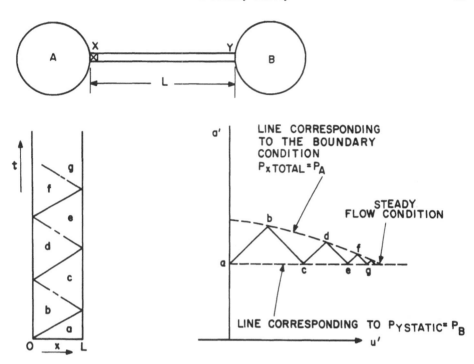

**Fig. 2–10** Transient flow in pipe connecting reservoirs A and B ($P_A > P_B$).

are generally identified as state diagrams. However, it is not normal practice to describe flow conditions in terms of $a'$ and $u'$ but rather in terms of pressure ratio, temperature ratio, and Mach number. These can all be established very easily from the $u' \sim a'$ plane. With reference to two arbitrary states 1 and 2:

$$\frac{P_2}{P_1} = \left(\frac{a'_2}{a'_1}\right)^{2\gamma/(\gamma-1)}, \qquad \frac{T_2}{T_1} = \left(\frac{a'_2}{a'_1}\right)^2, \qquad M_1 = \frac{u'_1}{a'_1}, \qquad M_2 = \frac{u'_2}{a'_2}$$

It can be seen that lines of constant Mach number can be added very easily, since they are straight lines, to $u' \sim a'$ diagrams. The pressure ratio relationship is restrictive since it is valid only for isentropic flow; however, isentropic flow is the only kind that can be analyzed using elementary theory. Before progressing to study the more general analytical method known as the method-of-characteristics it is worthwhile considering the shortcomings of the elementary treatment in more detail.

## 2.4  Shortcomings of the Elementary Theory

It is not, perhaps, surprising that there are major shortcomings in a theoretical treatment where a conscious effort has been made to maintain simplicity. It is

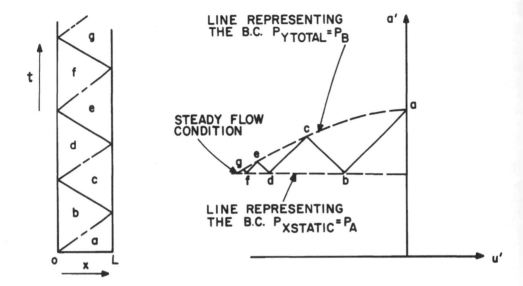

**Fig. 2–11** Transient flow for configuration shown in previous diagram when: $P_B > P_A$.

worthwhile examining these deficiencies to obtain an improved understanding of the physical aspects of nonsteady compressible flow and to justify the need for a more sophisticated theoretical treatment. Seven major shortcomings can be cited in the elementary treatment, namely an inability to cope with:

  i. nonuniform entropy,
 ii. tapered ducts,
iii. boundary conditions which vary with time in a continuous manner,
 iv. wave-intersection interactions,
  v. the fanning, or spreading, of expansion waves,
 vi. the bunching together of compression waves,
vii. discontinuous compression, or shock, waves.

Each of these will be dealt with in turn.

*2.4.1. Nonuniform entropy*

The exclusion of entropy variations within the flow field is endemic to the nature of the derivation of the $a \pm [(\gamma - 1)/2]u = \text{CONSTANT}$ relationship. This therefore restricts analysis to uniform entropy flows thereby ruling out the possibility of taking into account such factors as friction and heat transfer or initial entropy gradients or discontinuities.

With respect to entropy discontinuities it is worth mentioning that the reason why these are in general unacceptable, within the framework of an elementary approach, is because they give rise to wave reflections. However, the inherent possibility exists of what is termed impedance matching the fluids on each side of an entropy discontinuities. With impedance matched fluids an entropy discontinuity is transparent to an incident wave arriving at the discontinuity from either of the fluids. The reason why consideration was not given to this possibility earlier is that very special criteria must be met that limit the utility of the impedance matching concept. For example, for acoustic waves (i.e., isentropic waves of infinitesimal pressure ratio) the impedance matching condition is given by

$$\frac{\gamma_1}{a_1} = \frac{\gamma_2}{a_2}$$

where stations 1 and 2 are on opposite sides of the discontinuity. More comprehensive, but still restrictive, rules exist for the impedance matching of stronger waves. Impedance matching will be discussed in more detail in Chapter 5.

### 2.4.2  Tapered ducts

The general inability to handle flows in tapered channels is a shortcoming that is inherent to the concept of the elementary theory. In practice, the majority of real problems appear to relate to nonsteady compressible flows in essentially uniform area passages. Hence within the scope of an elementary treatment the restriction to flows in ducts of uniform cross-sectional area is not construed as particularly serious. In any case the influences of duct area change are incorporated fairly easily into computerized method-of-characteristics procedures.

The reason why it is not generally possible to encompass flows in ducts the cross-sectional area of which varies as continuous functions of flow path length is strongly related to the absence of rules for dealing with wave-on-wave interactions within the flow field. In particular cases, where sufficient boundary information is given, it is possible, within the framework of the elementary treatment, to handle flows in ducts with discontinuous area changes. A problem involving a discontinuous change of duct cross-sectional area is included in the exercise at the end of the chapter.

### 2.4.3  Time-dependent boundary conditions

It can be seen from the examples presented in Section 2.3, that the only forms of time-dependent boundary conditions considered were discontinuous functions of time as characterized by the instantaneous opening, or closing, of valves, etc. More general cases of boundary conditions varying with time in a continuous manner are not easily handled within the framework of the elementary treatment. This is in part a consequence of the absence of rules for dealing with wave-intersection interactions within the body of the flow field.

### 2.4.4  Wave-intersection interactions

There is an absence of rules for handling wave intersections as occur when waves propagating in opposite directions meet. When this happens the two waves appear, when viewed on the $x \sim t$ plane, to cross or intersect each other. It can be seen that this would be the case for the shock-tube flow illustrated in Fig. 2.9 if construction of the wave diagram were extended to a sufficiently large value of $t$. While the absence of rules for handling wave intersections limits the type, and sophistication, of problems that can be analyzed, it is not felt that this is of great practical importance within the limitations of a simple, introductory, theoretical treatment. Wave intersections can be handled rigorously and comprehensively by means of the method-of-characteristics, to be introduced in the next chapter, hence there seems little point in extending the elementary treatment to cover this topic.

### 2.4.5  Expansion wave fanning

The form of discontinuous expansion wave postulated in the elementary theoretical treatment is unrealistic as expansion waves, even those initiated as discontinuities, thicken, or fan out, as they propagate. The physical reason for the fanning phenomenon lies in the consideration that for an isentropic wave the local velocity of the wave front is the sum of the local acoustic and gas flow velocities, namely $a + u$. Hence for an expansion wave propagating to the right in fluid moving to the right, as illustrated in Fig. 2.12, all points on the wave move to the right with a velocity equal to the local sum of $a + u$. However, since the relationship

$$a - \left(\frac{\gamma - 1}{2}\right)u = \text{CONSTANT}$$

applies such a wave in addition to the isentropic relationship

$$\frac{a_2}{a_1} = \left(\frac{P_2}{P_1}\right)^{(\gamma - 1)/2\gamma}$$

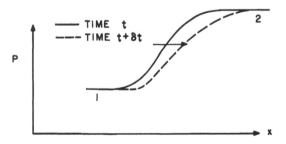

**Fig. 2–12** Expansion wave propagating to the right.

it can be seen that both $a$ and $u$ are greater at the top, or crest, of the wave than at the foot. Hence:

$$a_2 + u_2 > a_1 + u_1$$

and thus the crest of the wave propagates more rapidly than the foot. This, therefore, results in wave fanning. Further thought shows that fanning is a general characteristic of expansion waves not merely of the example studied. An approximation that can be used to establish the rate of propagation of an isentropic expansion wave, when the complexities of fanning are ignored, is to assume that the wave advances at a velocity that is the arithmetic average of the propagation velocities of the foot and the crest.

### 2.4.6   Isentropic compression waves

Similar reasoning to that used to explain the phenomenon of expansion wave fanning can be applied to initially nondiscontinuous compression waves. It can then be shown that such compression waves steepen with time. Figure 2.13 illustrates, on a pressure $\sim$ distance plane, a nonsteep isentropic compression wave propagating to the right in fluid moving to the right. A nonsteep compression wave, such as that of Fig. 2.13, propagates with a local velocity equal to the sum of the local acoustic and gas velocities, $a + u$ and since, as for the right propagating expansion wave:

$$a - \left(\frac{\gamma - 1}{2}\right)u = \text{CONSTANT}$$

also since

$$\frac{a_2}{a_1} = \left(\frac{P_2}{P_1}\right)^{(\gamma - 1)/2\gamma}$$

it can be seen that the crest of the wave advances more rapidly than the foot.

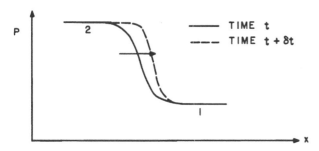

**Fig. 2–13** Compression wave propagating to the right.

Hence the wave will become, given a sufficient time, a discontinuity in the flow. Once this has occurred further consideration will show that it is no longer possible to satisfy, simultaneously, the three previously stated conditions relating to the propagation of an isentropic compression wave.

A discontinuous compression wave is termed a normal shock wave. A shock wave is irreversible hence, strictly, it is incorrect to treat a shock wave as isentropic. However, failure to consider a normal shock wave as such and, instead, to continue to apply isentropic relations does not produce serious quantitative errors provided the wave pressure ratios are relatively low, say 2:1 or less.

A similar approximation can be made to that suggested for expansion waves when estimating the rate of propagation of an isentropic compression wave, namely that the wave-propagation velocity is the arithmetic average of the propagation velocities of the wave foot and crest. This approximation can also be extended to weak normal shock waves where the two contributing propagation velocities are those that apply before the wave coalesces into a normal shock.

### 2.4.7  Shock waves

The theoretical pressure ratio across normal shock waves can be established precisely from well-known relationships the derivation of which can be found in most texts on compressible steady flow. These relations are based on applications of the mass, momentum, and energy conservation equations, gas law, etc., to flow in a stream tube of uniform cross-sectional area. Hence it can be shown that the static pressure ratio, $P_2/P_1$ of a normal shock wave is given by

$$\frac{P_2}{P_1} = \frac{2\gamma M_1^2 - (\gamma - 1)}{\gamma + 1} \tag{2.10}$$

where $M_1$ is the Mach number of the flow approaching the shock wave when it is stationary relative to the $x$ coordinate as shown in Fig. 2.14.

For the (hypothetical where discontinuous) isentropic compression wave:

$$a_1 + \left(\frac{\gamma - 1}{2}\right)u_1 = a_2 + \left(\frac{\gamma - 1}{2}\right)u_2$$

hence after introducing the definition of Mach number:

$$a_1\left[1 + \left(\frac{\gamma - 1}{2}\right)M_1\right] = a_2\left[1 + \left(\frac{\gamma - 1}{2}\right)M_2\right]$$

and thus

$$\frac{a_2}{a_1} = \left[\frac{1 + \dfrac{\gamma - 1}{2}M_1}{1 + \dfrac{\gamma - 1}{2}M_2}\right] \tag{2.11}$$

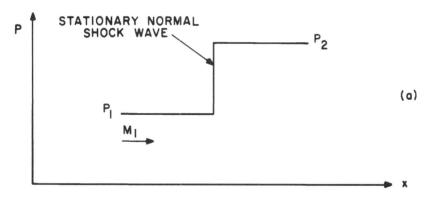

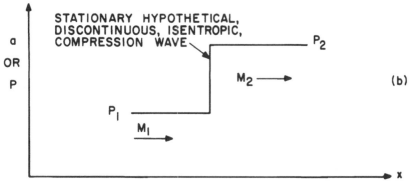

**Fig. 2–14** A stationary discontinuous compression wave.

after invoking classical isentropic relations Eq. (2.11) leads to the result:

$$\frac{P_2}{P_1} = \left[ \frac{1 + \dfrac{\gamma - 1}{2} M_1}{1 + \dfrac{\gamma - 1}{2} M_2} \right]^{2\gamma/(\gamma - 1)} \qquad (2.12)$$

When $M_1$ is chosen such that the wave is at rest as shown in Fig. 2.14 (a hypothetical situation since it implies a discontinuous, stationary, isentropic compression), then because the wave is isentropic, and at rest, and for these conditions only:

$$P_{\text{STAGNATION 1}} = P_{\text{STAGNATION 2}} \qquad (2.13)$$

Introducing the well-known steady-flow isentropic relation:

$$\frac{P_{\text{STAGNATION}}}{P} = \left[ 1 + \left( \frac{\gamma - 1}{2} \right) M^2 \right]^{\gamma/(\gamma - 1)} \qquad (2.14)$$

and combining Eqs. (2.13) and (2.14) leads to the result:

$$\frac{P_2}{P_1} = \left[ \frac{1 + \left(\dfrac{\gamma - 1}{2}\right)M_1^2}{1 + \left(\dfrac{\gamma - 1}{2}\right)M_2^2} \right]^{\gamma/(\gamma - 1)} \qquad (2.15)$$

Hence after equating the right-hand sides of Eqs. (2.12) and (2.15) an expression for $M_2$ in terms of $M_1$ can be obtained for the hypothetical discontinuous isentropic compression wave:

$$M_2 = \frac{4 - (3 - \gamma)M_1}{(3 - \gamma) + 2(\gamma - 1)M_1} \qquad (2.16)$$

Equation (2.16) can be solved, for selected values of $M_1$, and the corresponding values of $M_1$ and $M_2$ inserted in either Eq. (2.12) or (2.15) to permit $P_2/P_1$ to be evaluated. Figure 2.15 shows a presentation of such an evaluation for $\gamma = 1.4$. The results are compared with the corresponding values of $P_2/P_1$ for a normal shock. The latter data were obtained from Eq. (2.10). Figure 2.15 tends to support the assertion, made in the preceding subsection, that quantitative errors due to treating normal shocks as if they were isentropic are relatively small for shock (static) pressure ratios up to about 2:1. However, for significantly higher pressure ratios than this the error increases dramatically and it is clear that a normal shock wave must then be treated as such. Other aspects of normal shocks, for example evaluation of the associated entropy increase, will be considered specifically in Chapter 5.

### 2.4.8   Final comment

The restrictions imposed upon the elementary treatment imply that the accuracy of the solutions obtained, even for idealized situations without friction, heat transfer or variable entropy, decreases with increasing wave strength. The treatment is therefore of the asymptotic type that converges to quantitatively correct solutions, within the previously stated limitations, for infinitesimal pressure ratios.

The technique is particularly useful for providing qualitative and quantitative illustrations of some nonsteady flow phenomena with minimum effort. Use will be made later, in Chapter 7, of the elementary theory to identify, definitively and very easily, various pressure-exchanger cycles. Cyclic processes are represented by closed paths on $u' \sim a'$ (state) diagrams.

Analyses in greater depth than can be handled using the elementary theory usually require the introduction of the much more comprehensive method-of-characteristics approach.

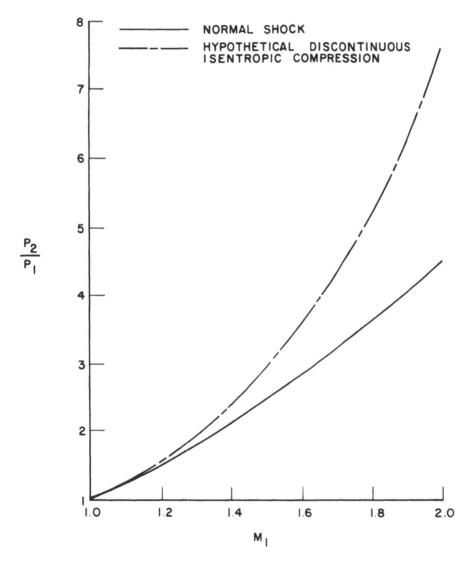

**Fig. 2–15** Comparison of static pressure ratios of stationary normal shock and stationary hypothetical isentropic compression wave.

**Problems**

NOTE. Assume, unless otherwise stated, that $\gamma = 1.4$.

2.1 Show that the expression

$$a \pm \left(\frac{\gamma - 1}{2}\right)u = \text{CONSTANT}$$

is valid for isentropic expansion waves.

2.2 Equate, over the time interval $\delta t$, the net momentum inflow into the control volume bounded by stations A and B of Fig. 2.3 to the momentum accumulation between A and B and hence show that the conservation of momentum principle is satisfied for an elemental isentropic compression wave.

2.3 Take, over the time interval $\delta t$ for the wave of Fig. 2.3, the sum of the net import of internal energy, flow work, and kinetic energy and equate to the accumulation of internal and kinetic energy with the control volume and hence show that an elemental isentropic compression wave satisfies the requirement that energy be conserved.

2.4 Compare the ratios of the initial and final pressures to the static pressure in the port, or valve, for an ideal isentropic filling process, similar to that shown in Fig. 2.7, when

    a. $\gamma = 1.6$

    b. $\gamma = 1.4$

    c. $\gamma = 1.3$

Assume $M = 0.5$ in each case.

2.5 An ideal isentropic emptying process, similar to that shown in Fig. 2.8, is intended to produce a Mach number of unity in the outlet port. What is the ratio of the initial container pressure to:

    a. the static pressure in the port?

    b. the stagnation pressure in the port?

and what is the ratio of the final container pressure to

    c. the static pressure in the port?

    d. the stagnation pressure in the port?

Assume that $\gamma = 1.4$.

2.6 Flow passes, uniformly, from left to right at a velocity of 100 m/s in a horizontal pipe of length 50 m. The pipe is connected to a large reservoir at the left-hand end and is open to the atmosphere at the right-hand end. The acoustic velocity within the pipe is 290 m/s. If the pressure is suddenly increased, slightly, in the reservoir how long is required for:

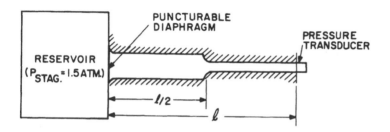

**Fig. 2–16** Diagram for problem 2.7.

    a. evidence of the pressure increase to be felt at the right-hand end of the pipe?

    b. a reflection from the open end to be transmitted to the upstream, left-hand end of the pipe?

2.7 Predict, for the stepped tube shown in Fig. 2.16, the pressure as a multiple of the reservoir stagnation pressure, which, following instantaneous puncture of the diaphragm, should be attained at the transducer immediately after wave reflection at the closed end. The entropy in the reservoir is equal to that in the stepped tube. The contraction is so shaped that flow through it can be assumed to be isentropic: Assume that the shock waves are sufficiently weak to be regarded as isentropic and that all other viscous effects are negligible. The pressure in the large reservoir is 1.5 atm, the initial pressure in the stepped tube is 1.0 atm.

    The area ratio of the contraction is such that the velocity of the flow approaching it is reduced by a factor of three after partial reflection, at the contraction, of the initial shock wave. Also determine the ratio of the maximum pressure attained at the closed end to that which would have been obtained had a tube of uniform cross-sectional area been used.

2.8 An inventor has proposed an intermittent constant volume combustor, as shown in Fig. 2.17, for use in gas turbines and other fuel burning devices. The purpose of the chamber is to produce a gain in pressure between the inlet and the outlet. In order to evaluate the idea a cold flow working model, also shown in Fig. 2.17, is to be constructed. In the model the combustion process has been replaced by air addition via valve C. Valves A and B operate instantaneously and are either fully open or fully shut.

    The sequence of operation of the valves is:

    i. with valves A and B shut, valve C opens to permit compressed air inflow,

    ii. valve C is shut,

    iii. valve B is opened for sufficient time to allow the minimum possible pressure to be achieved in length $l$; valve B is then shut,

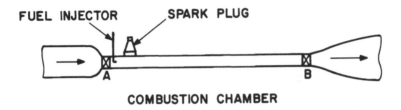

COMBUSTION CHAMBER

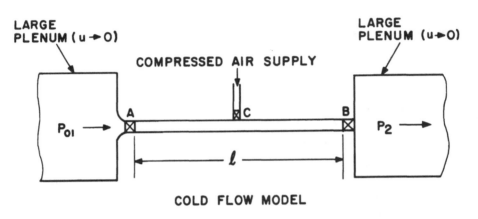

COLD FLOW MODEL

**Fig. 2–17** Diagram for problem 2.8.

iv. valve A is opened for sufficient time to allow the highest possible pressure to be achieved in the length $l$; valve A is then shut.

The sequence (i) through (iv) is then repeated, etc.

Evaluate, on a $u \sim a$ basis, the pressure rise as a multiple of the inlet stagnation pressure, $P_{o1}$, to which length $l$ must be subjected due to inflow through valve C, when $P_2/P_{o1} = 1.2$. The maximum flow Mach number in length $l = 0.5$. What would be the pressure ratio across the device if the dynamic head at valve B could be recovered?

# 3

## METHOD-OF-CHARACTERISTICS

The method-of-characteristics as employed in the analysis of one-dimensional nonsteady flows, or two-dimensional steady supersonic flows, constitutes a graphical or numerical procedure for solving the so-called *hyperbolic* partial differential equations describing the flow field. The name of the method is derived from the solution procedure, or method, that involves following particular, or characteristic, paths in the flow field. These paths can be shown to be those on which discontinuities may exist in the solution domain. The possibility of discontinuities, as represented for example by pressure waves in nonsteady fluid flows or in steady supersonic flows, is a significant feature of the solution of partial differential equations of the hyperbolic type.

### 3.1 Types of Partial Differential Equations

Second-order partial differential equations are classified as either elliptic, hyperbolic, or parabolic.

Consider the general second-order partial differential equation for the case of two independent variables $x$ and $y$:

$$A \frac{\partial^2 \phi}{\partial x^2} + 2B \frac{\partial^2 \phi}{\partial x \, \partial y} + C \frac{\partial^2 \phi}{\partial y^2} = D$$

where the coefficients $A$, $B$, $C$, and $D$ can be functions of $x$, $y$, $\partial \phi / \partial x$, and $\partial \phi / \partial y$. It can be shown that the type of solution obtained depends upon whether $AC - B^2$ is greater than, less than, or equal to zero. Each possibility will be considered in turn.

$AC - B^2 > 0$

Partial differential equations characterized by this relationship are said to be of the *elliptic* type. The term "elliptic" is used on the basis of an analogy with the general second-order equation identifying conic sections, namely:

$$ax^2 + 2bxy + cy^2 = d$$

where $a$, $b$, $c$, and $d$ are constants. The particular type of conic section

33

represented can be shown to be

| | | |
|---|---|---|
| an ellipse | when | $ac - b^2 > 0$ |
| a hyperbola | when | $ac - b^2 < 0$ |
| a parabola | when | $ac - b^2 = 0$ |

In order to solve partial differential equations of the elliptic type it is essential to specify conditions on a boundary enclosing the region within which a solution is required. A feature of the type of solution obtained is that the influence of, for example, a local change of boundary condition is felt everywhere within the enclosed region. An example of an equation of the elliptic type is the Laplace equation.

## $AC - B^2 < 0$

This condition identifies *hyperbolic* partial differential equations. For differential equations of this type the solution procedure (in two-dimensions) identifies distinct families of characteristics and a feature of the solution is that irregularities in $\phi$ are not necessarily smoothed out but can persist and be transferred from place to place (on the $x \sim y$ plane as implied by the general equation, or on the $x \sim t$ plane if $t$ replaces $y$) in the region in which the solution is obtained.

It is not generally possible, that is assuming that the solution is not known in advance, to specify boundary conditions that enclose the region; however, certain boundary conditions must be specified.

As stated earlier, examples of hyperbolic equations are those defining two-dimensional steady supersonic flow (on the $x \sim y$ plane) and one-dimensional nonsteady flow (on the $x \sim t$ plane).

## $AC - B^2 = 0$

This characterizes partial differential equations of the *parabolic* type. Examples of parabolic differential equations are sometimes found in mathematical models of boundary layers. A general feature is a tendency to smooth out gradients of $\phi$ within the region in which the solution is being obtained when localized perturbations are made to the boundary conditions.

As a matter of practical importance it should be noted that when $A$, $B$, and $C$ are not constants the classification of a differential equation may change within the region within which a solution is required. As an example, for a steady two-dimensional accelerating flow the partial differential equations describing the flow will change from elliptic, when the flow can be treated as incompressible and subsonic, to hyperbolic in the supersonic portion of the stream.

The foregoing notes are not, of course, offered as an explanation of the theory of the classification of differential equations merely as a convenient description of the outcome of such classification. Rigorous descriptions of the theory involved are available (Courant and Friedrichs, 1948; Sneddon, 1957).

Physically based descriptions of the more relevant aspects of the theory have been presented by Shapiro (1953) and by Riley (1974).

Before focusing attention on the main task of this chapter, that of presenting a derivation of the equations necessary to implement the method-of-characteristics solution procedure, it is important to list the assumptions involved.

## 3.2   Assumptions

The following assumptions are implicit in the analysis presented in the next section:

   i. flow is one-dimensional,

  ii. flow is nonsteady (i.e., time dependent),

 iii. flow is compressible,

 iv. variations of entropy are taken into account,

  v. wall friction within the flow channel is taken into account by means of a friction force (boundary layer not identified as such; consistent with one-dimensional flow),

 vi. heat transfer from, or to, duct walls taken into account (temperature gradients lateral to the axis of the channel not identified; consistent with one-dimensional flow),

 vii. thermal influences of exothermic or endothermic reactions taken into account as for heat transfer,

viii. flow-channel cross-sectional area can vary as a function of $x$ (only) within limitations consistent with a one-dimensional flow model,

 ix. ratio of specific heats, $\gamma$, invariant.

It should be noted that assumptions (v), (vi), and (vii) do not imply that friction, heat transfer, and heat release/absorption rates must be constants; each may vary as functions of both $x$ and $t$. Assumption (ix) is consistent with the basic concept of the method-of-characteristics solution procedure. However, variations of $\gamma$ can be accommodated by making additional provisions for this circumstance. These are, in practice, easiest to implement in numerical solutions.

## 3.3   Derivation of the Differential Equations

These will be derived from first principles by a sequence of algebraic manipulations and substitutions performed on the basic conservation equations. In the overall sense the derivation is classical in nature; however, the details of the procedure follow fairly closely, but not exactly, those of Spalding (1969). The aim is to set up comprehensive differential equations as a basis for both graphical and numerical solution procedures.

### 3.3.1   Conservation equations

Consideration of time-dependent, one-dimensional compressible flow within a

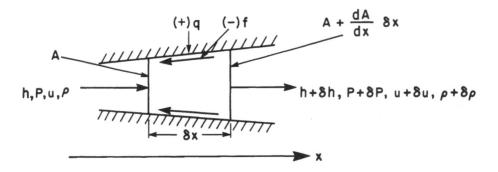

**Fig. 3–1** Representative fluid element in a duct of nonuniform cross-sectional area.

channel, as shown in Fig. 3.1, of nonuniform cross-sectional area when friction and heat transfer are present gives for the:

conservation of mass (continuity)

$$\frac{\partial \rho}{\partial t} + \frac{\partial}{\partial x}(\rho u) + \frac{\rho u}{A}\frac{dA}{dx} = 0 \qquad (3.1)$$

conservation of momentum

$$\frac{\partial}{\partial t}(\rho u) + \frac{\partial}{\partial x}(\rho u^2 + P) + \frac{\rho u^2}{A}\frac{dA}{dx} = -f\rho \qquad (3.2)$$

conservation of energy

$$\frac{\partial}{\partial t}\left\{\rho\left[h - \frac{P}{\rho} + \frac{u^2}{2}\right]\right\} + \frac{\partial}{\partial x}\left\{\rho u\left(h + \frac{u^2}{2}\right)\right\} + \frac{1}{A}\left\{\rho u\left(h + \frac{u^2}{2}\right)\right\}\frac{dA}{dx} = q\rho \qquad (3.3)$$

where:

$h \equiv$ enthalpy,

$f \equiv$ friction force at the duct wall per *unit mass* of fluid (positive when acting on the fluid in the direction of $x$ increasing),

$q \equiv$ heat transfer rate through the duct wall *into* fluid per *unit mass* of fluid and/or heat release rate per *unit mass* of fluid.

*Note:* The derivation can be carried out where $f$ and $q$ are defined such as to refer to either a unit mass (Rudinger, 1969) or a unit volume (Spalding, 1969) of fluid. The final results of each derivation can readily be converted into the form of the other. Here it was decided to refer $f$ and $q$ to a unit mass of fluid since this simplifies the analysis very slightly

### 3.3.2   Rearranged conservation equations

The conservation equations can now be rearranged in order to make their presentation more uniform and systematic. This in turn facilitates the later introduction of a differential operator that simplifies subsequent manipulations.

Carrying out the differentiation implied in the second term of (3.1) yields:

conservation of mass

$$\frac{\partial \rho}{\partial t} + u \frac{\partial \rho}{\partial x} + \rho \frac{\partial u}{\partial x} = -\frac{\rho u}{A} \frac{dA}{dx} \tag{3.4}$$

similar reorganization of (3.2) gives, after substitution from (3.4):

conservation of momentum

$$\frac{\partial u}{\partial t} + u \frac{\partial u}{\partial x} + \frac{1}{\rho} \frac{\partial P}{\partial x} = -f \tag{3.5}$$

reorganization of (3.3) yields, after substitutions from (3.4) and (3.5):

conservation of energy

$$\frac{\partial h}{\partial t} + u \frac{\partial h}{\partial x} - \frac{1}{\rho} \left\{ \frac{\partial P}{\partial t} + u \frac{\partial P}{\partial x} \right\} = q + uf \tag{3.6}$$

Inspection of Eqs. (3.4) to (3.6) inclusive reveals a uniformity of the required type in the first pair of terms on the L.H.S. and, in the case of Eq. (3.6) only, a similar pattern in the second pair of terms on the L.H.S. also.

### 3.3.3   Introduction of the Substantial Derivative

Defining a differential operator, $S$, and substituting it into the conservation equations allows their presentation to be simplified.

Defining the operator thus:

$$S \equiv \frac{\partial}{\partial t} + u \frac{\partial}{\partial x} \tag{3.7}$$

which is commonly termed the Substantial Derivative, and substituting it into the appropriate terms of Eqs. (3.4) to (3.6) inclusive allows the following results to be obtained:

conservation of mass

$$S\rho + \rho \frac{\partial u}{\partial x} = -\frac{\rho u}{A} \frac{dA}{dx} \tag{3.8}$$

conservation of momentum

$$Su + \frac{1}{\rho} \frac{\partial P}{\partial x} = -f \tag{3.9}$$

conservation of energy

$$Sh - \frac{1}{\rho} SP = q + uf \qquad (3.10)$$

In order to proceed further it is first necessary to eliminate, from the L.H.S. of Eqs. (3.8) to (3.10) inclusive, the dependent variables, $P$, $\rho$, and $h$ in favor of the acoustic velocity, $a$, and entropy $s$. The thermodynamic relationships required for this task are given in Section 3.3.4.

### 3.3.4   Thermodynamic relations

Ideal gas law:

$$\frac{P}{\rho} = RT \qquad (3.11)$$

from logarithmic differentiation of (3.11)

$$\frac{dP}{P} - \frac{d\rho}{\rho} = \frac{dT}{T} \qquad (3.12)$$

From the First Law of Thermodynamics applied along a reversible path connecting two adjacent, but arbitrarily chosen, state points:

$$T\,ds = dh - \frac{dP}{\rho} \qquad (3.13)$$

The expression for the acoustic velocity:

$$a = \sqrt{\gamma RT} \qquad (3.14)$$

differentiating (3.14)

$$2\frac{da}{a} = \frac{dT}{T} \qquad (3.15)$$

Expressing enthalpy in terms of $c_p$ and $T$ (i.e., $h = c_p T$) and differentiating:

$$dh = c_p\,dT \qquad (3.16)$$

These expressions allow others to be derived that ultimately permit the required substitutions for $P$, $\rho$, and $h$ to be made in Eqs. (3.8), (3.9), and (3.10).

### 3.3.5   Derived relationships

The following expressions were obtained by manipulation of Eqs. (3.11) to (3.16) inclusive:

$$\frac{d\rho}{\rho} = \left(\frac{2}{\gamma - 1}\right)\frac{da}{a} - \frac{ds}{R} \qquad (3.17)$$

$$\frac{dP}{P} = \left(\frac{2\gamma}{\gamma - 1}\right)\frac{da}{a} - \frac{ds}{R} \tag{3.18}$$

$$dh - \frac{dP}{\rho} = \frac{1}{\gamma}\frac{a^2}{R}ds \tag{3.19}$$

It is now possible to proceed to make the required changes of variable in Eqs. (3.8), (3.9), and (3.10).

### 3.3.6 Resultant forms of the conservation equations

The second (momentum) conservation equation is obtained from substitution in (3.9) from (3.18):
conservation of momentum

$$S\left(\frac{\gamma - 1}{2}u\right) + a\frac{\partial a}{\partial x} = \frac{a^2}{2c_p}\frac{\partial s}{\partial x} - \left(\frac{\gamma - 1}{2}\right)f \tag{3.20}$$

The final form of the energy equation is obtained from substitution from (3.19) in (3.10):
conservation of energy

$$Ss = \frac{\gamma R}{a^2}(q + uf) \tag{3.21}$$

The continuity equation is obtained from substitutions from (3.17) and (3.21) in (3.8) thus:

conservation of mass

$$Sa + a\frac{\partial}{\partial x}\left(\frac{\gamma - 1}{2}u\right) = \left(\frac{\gamma - 1}{2}\right)\left\{\frac{\gamma}{a}(q + uf) - a\frac{u}{A}\frac{dA}{dx}\right\} \tag{3.22}$$

It is now possible, after introducing the Riemann variables, to develop the final differential equations.

### 3.3.7 Introduction of the Riemann variables

It is found that more convenient forms of (3.20) and (3.22) can be obtained if the S operator is replaced by new differential operators and the variables a and u are combined on the L.H.S. of those equations into new variable, m and n, that are each combinations of a and u. The variables m and n are known as the Riemann variables.

The new differential operators are:

$$M \equiv S - a\frac{\partial}{\partial x} \tag{3.23}$$

and

$$N \equiv S + a\frac{\partial}{\partial x} \tag{3.24}$$

By invoking the definition of $S$ [Eq. (3.7)] $M$ and $N$ can be written as:

$$M = \frac{\partial}{\partial t} + (u - a)\frac{\partial}{\partial x} \tag{3.25}$$

$$N = \frac{\partial}{\partial t} + (u + a)\frac{\partial}{\partial x} \tag{3.26}$$

The physical meaning of the operators $M$, $N$, and $S$ can be explained quite simply. $M\phi$ can be interpreted as the variation with time of the variable ($\phi$ in this case) along a line, or characteristic, of slope $1/(u - a)$ on the $x \sim t$ plane. This point is clarified by considering, separately, each term on the R.H.S. of Eq. (3.25). The first term implies the change in the variable quantity ($\phi$) per unit time with $x$ constant. The second term implies the change of $\phi$ with $x$ multiplied by the distance traveled per unit time in the $x$ direction.

Similarly $N\phi$ implies the variation of $\phi$ with time along a line of slope $1/(u + a)$ on the $x \sim t$ plane.

For the case of the $S$ operator (Substantial Derivative) the meaning of $S\phi$ is the variation of $\phi$ with time along a line of slope $1/u$ on the $x \sim t$ plane. Figure 3.2 illustrates this situation, on the $x \sim t$ plane, by considering the change of $\phi$ between stations 1 and 2 which are joined by a line having a slope, $\delta t/\delta x$, of $1/u$.

The Riemann variables on which $M$ and $N$ will operate are defined thus:

$$m \equiv a - \left(\frac{\gamma - 1}{2}\right)u \tag{3.27}$$

$$n \equiv a + \left(\frac{\gamma - 1}{2}\right)u \tag{3.28}$$

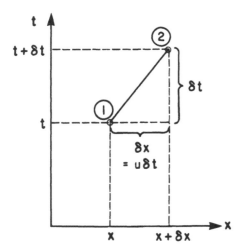

THE VARIATION-WITH-TIME OF $\phi$ BETWEEN ① AND ② $= \delta\phi$ WHERE:

$$\delta\phi = S\phi\delta t = \left[\frac{\partial\phi}{\partial t} + \frac{u\partial\phi}{\partial x}\right]\delta t$$

**Fig. 3–2** Interpretation of the Substantial Derivative.

It is also convenient, in order to simplify the final form of the differential equations, to define the following variables relating to terms on the R.H.S. of Eqs. (3.20) to (3.22) inclusive:

$$E \equiv \left(\frac{\gamma - 1}{2}\right)\left[\frac{\gamma}{a}(q + uf) - a\frac{u}{A}\frac{dA}{dx}\right] \tag{3.29}$$

$$F \equiv \frac{a^2}{2c_p} \tag{3.30}$$

$$G \equiv \frac{\gamma R}{a^2}(q + uf) \tag{3.31}$$

$$H \equiv \left(\frac{\gamma - 1}{2}\right)f \tag{3.32}$$

Likewise it is convenient to define terms representative of the three slopes $1/(u - a)$, $1/(u + a)$, and $1/u$ on the $x \sim t$ plane:

$$\beta_m \equiv (u - a) \tag{3.33}$$

$$\beta_n \equiv (u + a) \tag{3.34}$$

$$\beta_s \equiv u \tag{3.35}$$

*Note:* The $\beta$ variables are the reciprocal of the slopes of the three lines, or characteristics, on the $x \sim t$ plane.

### 3.3.8 Final partial differential equations

From Eqs. (3.20), (3.22), the definitions of $M$ and $N$, Eqs. (3.27) to (3.30) inclusive and Eq. (3.32):

$$Mm = E - F\frac{\partial s}{\partial x} + H \tag{3.36}$$

$$Nn = E + F\frac{\partial s}{\partial x} - H \tag{3.37}$$

and from Eqs. (3.21) and (3.31):

$$Ss = G \tag{3.38}$$

At this point it is worth pausing to consider the physical significance of Eqs. (3.36), (3.37), and (3.38); these equations are those describing the three characteristic families, $m$, $n$, and $s$, on the $x \sim t$ plane.

Equation (3.36) shows the variation of $m$ along a line of slope $1/\beta_m$ on the $x \sim t$ plane depends only upon friction ($f$), heat transfer rate ($q$), entropy gradient ($\partial s/\partial x$), and the duct-area variation term, $(1/A)(dA/dx)$. Thus it can be seen that:

i. for isentropic flow everywhere in a duct of uniform area

$$\left( f = q = \frac{\partial s}{\partial x} = \frac{dA}{dx} = 0 \right)$$

$m$ is invariant along an $m$ characteristic, i.e., a line of slope $1/\beta_m$ on the $x \sim t$ plane,

ii. for subsonic flow (cases of most interest) the slope $(1/\beta_m = 1/(u - a)$ of an $m$ characteristic is always negative on the $x \sim t$ plane.

Similarly Eq. (3.37) shows that:

iii. for isentropic flow everywhere in a duct of uniform area $n$ is invariant along an $n$ characteristic, i.e., a line of slope $1/\beta_n$ on the $x \sim t$ plane,

iv. for subsonic flow the slope of an $n$ characteristic, $[1/\beta_n = 1/(u + a)]$ is always positive on the $x \sim t$ plane.

From Eq. (3.38) it can be deduced that:

v. when friction and heat transfer are absent, entropy $s$ is invariant along an $s$ characteristic. This in itself does not necessarily imply isentropic flow in the flow field as a whole, i.e., homentropic flow,

vi. if the flow is homentropic, $s$ is invariant in the entire region and not merely along an $s$ characteristic,

vii. the slope of the $s$ characteristic $(1/\beta_s = 1/u)$ is such that $s$ characteristics lie along particle paths,

viii. the slope of an $s$ characteristic can be positive or negative (i.e., as $u$ is positive or negative).

A factor related to the first pair of the three final equations, namely, Eqs. (3.36) and (3.37) that still requires explanation is the appearance of the partial derivative term $\partial s/\partial x$ in these equations. This arises from the consideration that entropy can only decrease or increase along a particle path due to the combined influences of friction and heat transfer; friction and heat transfer are fully taken care of by the $E$ and $H$ terms. However, $s$ will vary with $x$ if the flow field is of nonuniform entropy; the $\partial s/\partial x$ terms take account of this. A flow field may achieve a condition of nonuniform entropy due to earlier influences of heat transfer and friction or due to the addition, to the duct in which the unsteady flow is occurring, of fluid of nonuniform entropy.

Having explained in a physical sense the meaning of Eqs. (3.36), (3.37), and (3.38) it is now appropriate to sketch, $m$, $n$, and $s$ characteristics on the $x \sim t$ plane. Individual $m$, $n$, and $s$ characteristics are shown in Fig. 3.3. for subsonic flow with positive $u$ and for subsonic flow with a negative value of $u$. It should be noted that the characteristics depicted in Fig. 3.3 are not those applicable to a unique location on the $x \sim t$ plane. If they applied to a single location they

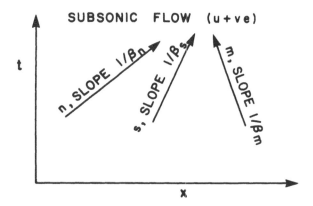

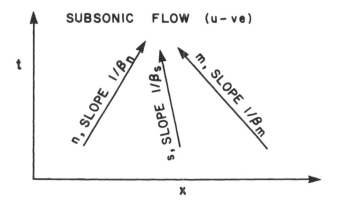

**Fig. 3–3** Members of the three characteristic families.

would all intersect at that location. Figure 3.4 is a sketch of the $m$, $n$, and $s$ characteristics for a particular point, labeled $A$, on an $x \sim t$ plane for a subsonic flow with positive $u$. The characteristics will not necessarily be straight lines, as indicated in Fig. 3.4.

Although Eqs. (3.36) to (3.38) inclusive cover the general case, within the limitations set forth in Section 3.2, very considerable simplifications are possible for the special case of homentropic flow in a duct of uniform area.

### 3.3.9  Simplified equations for homentropic flow

When flow is isentropic everywhere within the flow field (i.e., homentropic flow) in a duct of uniform cross-sectional area equations (3.36), (3.37), and (3.38) reduce to two very simple results.

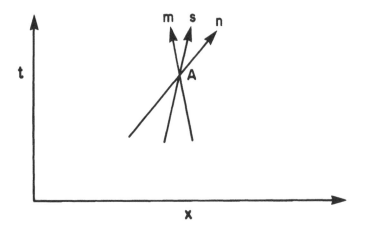

**Fig. 3–4** The three characteristics passing through a typical point, A, on the $x \sim t$ plane.

Equation (3.38) is redundant for homentropic flow because $s$ is constant for all $x$ and $t$. Equations (3.36) and (3.37) reduce, respectively, to:

$$Mm = 0 \tag{3.39}$$

$$Nn = 0 \tag{3.40}$$

or, by invoking the definitions of $m$ and $n$ and the meaning of $Mm$ and $Nn$, since according to Eqs. (3.39) and (3.40) there is no change of $m$, or $n$, along the $m$, or $n$, characteristic paths on the $x \sim t$ plane:

$$a - \left(\frac{\gamma - 1}{2}\right)u = \text{CONSTANT} \tag{3.41}$$

along an $m$ characteristic and

$$a + \left(\frac{\gamma - 1}{2}\right)u = \text{CONSTANT} \tag{3.42}$$

along an $n$ characteristic.

Equations (3.41) and (3.42) are similar to results obtained earlier with a more elementary analysis and should be compared with Eq. (2.9) of the previous chapter. Here, however, the implication is that $m$ and $n$ characteristics can both exist simultaneously and may, therefore, intersect each other on the $x \sim t$ plane. It can also be seen, from the expressions for $\beta_m$ and $\beta_n$, that their slopes on the $x \sim t$ plane will be modified mutually when the characteristics intersect. Thus a series of discrete $m$ and $n$ characteristics on the $x \sim t$ plane could appear as sketched in Fig. 3.5.

Due to the nature of Eqs. (3.41) and (3.42) the characteristics of Fig. 3.5 constitute, when presented on the $u \sim a$ plane, a series of intersecting straight

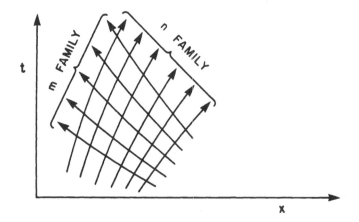

**Fig. 3–5** A hypothetical mesh of m and n characteristics describing an isentropic flow field.

lines all with slopes of $\pm 45°$ provided $a$ is plotted to a scale of $2/(\gamma - 1)$ times that of $u$. Figure 3.6 is a sketch of the probable appearance of the characteristics on a $u \sim a$ plane where the $a$ scale has been chosen to be $2/(\gamma - 1)$ times that used for $u$. Figure 3.6 should be compared with Fig. 2.6.

In order to proceed further it is necessary to consider the details of methods for solving Eqs. (3.39) and (3.40) and, for nonisentropic cases, Eqs. (3.36) to (3.38). Before doing this it is desirable and convenient to convert all the variables into dimensionless, normalized forms.

### 3.3.10  Normalization of equations

The normalized variables are distinguished from their unnormalized brethren by a prime. The normalizing factors were chosen such that they are both convenient and consistent. The normalized variables are:

**Variables**

$$\frac{dA'}{dx'} \equiv \frac{l}{A}\frac{dA}{dx} \quad \text{(Note: } dA' \equiv dA/A) \tag{3.43}$$

$$a' \equiv a/a_{\text{REF}} \tag{3.44}$$

$$f' \equiv \gamma f l/a_{\text{REF}}^2 \tag{3.45}$$

$$m' \equiv m/a_{\text{REF}} \tag{3.46}$$

$$n' \equiv n/a_{\text{REF}} \tag{3.47}$$

$$q' \equiv \gamma q l/a_{\text{REF}}^3 \tag{3.48}$$

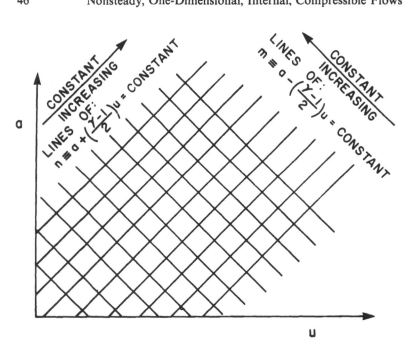

**Fig. 3–6** Representative state ($a \sim u$) diagram for an isentropic flow field.

$$s' \equiv s/R \qquad \text{(when: } s'_{REF} = 0) \qquad\qquad (3.49)$$

$$t' \equiv ta_{REF}/l \qquad\qquad (3.50)$$

$$x' \equiv x/l \qquad\qquad (3.51)$$

$$u' \equiv u/a_{REF} \qquad\qquad (3.52)$$

**Operators**

$$M' \equiv (l/a_{REF})M \qquad\qquad (3.53)$$

$$N' \equiv (l/a_{REF})N \qquad\qquad (3.54)$$

$$S' \equiv (l/a_{REF})S \qquad\qquad (3.55)$$

*Note:* Definitions (3.44) and (3.52) are identical to the definitions for normalized acoustic and particle velocities introduced in Chapter 2. It is also worth noting that the dimensionless time, $t'$, is defined in such a way that a single unit of dimensionless time corresponds to the time taken for an acoustic wave, traveling at $a_{REF}$, to translate through a distance $l$. The definition of $x'$ is such that $x'$ is merely a fraction of the reference length $l$.

Invoking the normalized variables allows Eqs. (3.29) to (3.35) to be rewritten thus:

$$E' = \left(\frac{\gamma - 1}{2}\right)\left[\left(\frac{q' + u'f'}{a'}\right) - a'u'\frac{dA'}{dx'}\right]$$                      (3.56)

$$F' = \left(\frac{\gamma - 1}{2\gamma}\right)a'^2$$                      (3.57)

$$G' = \left(\frac{q' + u'f'}{a'^2}\right)$$                      (3.58)

$$H' = \left(\frac{\gamma - 1}{2\gamma}\right)f'$$                      (3.59)

$$\beta'_m = u' - a'$$                      (3.60)

$$\beta'_n = u' + a'$$                      (3.61)

$$\beta'_s = u'$$                      (3.62)

And the normalized forms of the differential Eqs. (3.36) to (3.38) are

$$M'm' = E' - F'\frac{\partial s'}{\partial x'} + H'$$                      (3.63)

$$N'n' = E' + F'\frac{\partial s'}{\partial x'} - H'$$                      (3.64)

$$S's' = G'$$                      (3.65)

Similarly, normalized forms of Eqs. (3.41) and (3.42), for homentropic flow, are

$$a' - \left(\frac{\gamma - 1}{2}\right)u' = \text{CONSTANT}$$                      (3.66)

along an $m'$ characteristic and

$$a' + \left(\frac{\gamma - 1}{2}\right)u' = \text{CONSTANT}$$                      (3.67)

### 3.3.11   Derived quantities

The following quantities relating the main normalized method-of-characteristics variables ($m'$, $n'$, and $s'$) to other physically meaningful ones are of use in converting the results of method-of-characteristics calculations back into familiar terms.

From Eqs. (3.27), (3.28), (3.43), (3.46), (3.47), and (3.52):

$$a' = \frac{m' + n'}{2}$$                      (3.68)

$$u' = \frac{n' - m'}{\gamma - 1}$$                      (3.69)

also:

$$\text{Mach number} = u'/a' \tag{3.70}$$

From thermodynamic relations where $T' \equiv T/T_{REF}$:

$$T' = (a')^2 \tag{3.71}$$

An expression for $P'$ $(\equiv P/P_{REF})$ in terms of $a'$ and $s'$ is also needed. Writing $dh = c_p\, dT$, invoking the gas law $P/\rho = RT$, and substituting from these in Eq. (3.13):

$$ds = c_p \frac{dT}{T} - R \frac{dP}{P}$$

thus for an ideal gas, where $c_p$, is constant, integration gives

$$s - s_{REF} = c_p \ln\left(\frac{T}{T_{REF}}\right) - R \ln\left(\frac{P}{P_{REF}}\right)$$

or

$$\frac{s - s_{REF}}{R} = \left(\frac{\gamma}{\gamma - 1}\right) \ln\left(\frac{T}{T_{REF}}\right) - \ln\left(\frac{P}{P_{REF}}\right)$$

and thus

$$\frac{s - s_{REF}}{e^R} = \left(\frac{T}{T_{REF}}\right)^{\gamma/(\gamma - 1)} \frac{P_{REF}}{P}$$

since $s_{REF} \equiv 0$ and $T/T_{REF} \equiv T'$, invoking Eq. (3.71) and substituting for $s/R$ in terms of $s'$ from Eq. (3.49) gives

$$P' = (a')^{2\gamma/(\gamma - 1)} e^{-s'} \tag{3.72}$$

It should be noted that Eq. (3.72) implies that $P_{REF}$ and $T_{REF}$ (and hence $a_{REF}$) are defined in a consistent manner as indicated in Fig. 3.7. For an isentropic case where $s' = s'_{REF}$ and since $s'_{REF} \equiv 0$ (see Eq. 3.49), Eq. (3.72) reduces to:

$$P' = (a')^{2\gamma/(\gamma - 1)} \tag{3.73}$$

this expression corresponds to one presented in Chapter 2 for isentropic flow.

In order to proceed to obtain full solutions for most nonsteady compressible flow problems using the method-of-characteristics it is essential to define suitable boundary conditions. However, demonstrations can now be carried out to illustrate a noncomputerized solution procedure applicable within the body of the flow field in a duct of uniform cross-sectional area when friction, heat transfer, and entropy gradients are absent.

### 3.4  Illustrative Examples

Two examples, featuring homentropic flow, will be considered. In both cases solutions will be obtained graphically. The first example involves only a single

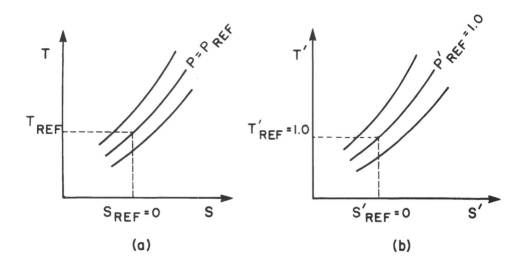

**Fig. 3–7** Self-consistent reference conditions (a) without normalization, (b) normalized presentation.

family of characteristics representing a compression wave. In the second example two families of characteristics are manipulated simultaneously to show what happens when two waves moving in opposite directions intersect.

### 3.4.1   Example with one family of characteristics

Consider the isentropic wave, shown in Fig. 3.8, represented by an $n'$ value of 1.1. The wave moves to the right in gas initially at rest in a duct of uniform cross-sectional area. Predict when and where the wave will steepen into a normal shock; $m'$ is unity upstream of the wave (i.e., in front of the wave), $\gamma = 1.4$ throughout the flow field.

*Solution:*

From Eq. (3.69), since $u' = 0$ in front of the wave; $m' = n'$; hence, because $m' = 1.0$, $n' = 1.0$.

   Since the wave is moving to the right it is represented by $n'$ characteristics; further, since there are no waves propagating to the left there is no variation of $m'$ within the flow field (changes in $m'$ can only be induced by $m'$ characteristics), thus $m'$ is also unity behind the wave; i.e., $m' = 1.0$ everywhere within the flow field.

   Furthermore, from (3.69):

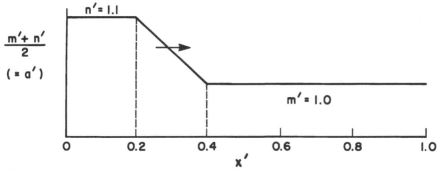

**Fig. 3–8** Initial conditions for single-wave example.

$$u'_{\text{Behind Wave}} = \left(\frac{n' - m'}{\gamma - 1}\right)_{\text{Behind Wave}}$$

$$= \frac{1.1 - 1.0}{0.4} = 0.25$$

also from (3.68)

$$a'_{\text{Behind Wave}} = \left(\frac{m' + n'}{2}\right)_{\text{Behind Wave}}$$

$$= \frac{1.0 + 1.1}{2} = 1.05$$

The $u' \sim a'$ diagram can now be constructed. This is presented in Fig. 3.9.

Figure 3.10, the $x' \sim t'$ diagram, can then be produced showing the progress of the wave. The slope, on the $x' \sim t'$ plane, of the characteristic representing the foot of the wave is that of an $n' = 1.0$ characteristic intersecting with an $m' = 1.0$ characteristic.

$$\text{Since slope} = \frac{1}{\beta_n} = \frac{1}{u' + a'}$$

from the $u' \sim a'$ plane (Fig. 3.9):

$$\text{slope} = \frac{1}{0 + 1.0} = 1.0$$

Similarly, the slope of the characteristic representing the wave crest is, from the $u' \sim a'$ plane:

$$\frac{1}{0.25 + 1.05} = 0.77$$

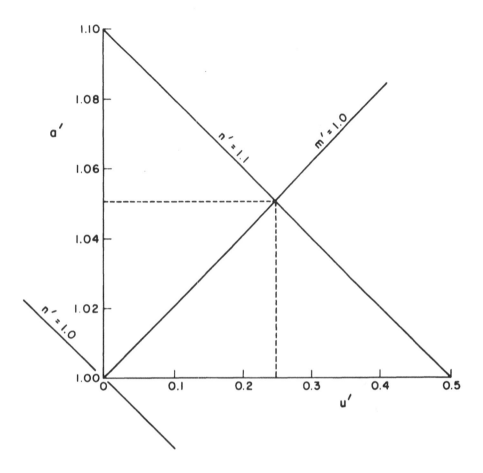

**Fig. 3–9** State diagram for single-wave example.

Hence from the $x' \sim t'$ diagram it can be seen that the wave steepens into a normal shock wave when:

$$x' = 1.05, \qquad t' = 0.66$$

This is the required solution.

### 3.4.2 Example with two families of characteristics

This example involves the interaction of the two families of characteristics in an isentropic flow field within a duct of uniform cross-sectional area.

Consider two isentropic waves approaching each other as shown in Fig. 3.11. The undisturbed fluid between the waves is at rest.

Predict the profile on the $P'^{(\gamma - 1)/2\gamma}$ versus $x'$ plane at time $\Delta t' = 0.4$ later

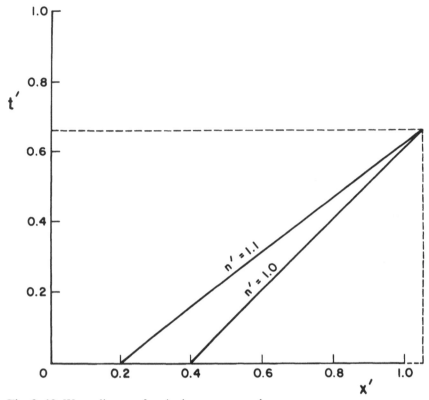

**Fig. 3–10** Wave diagram for single-wave example.

than for the initial situation depicted in Fig. 3.11. Assume $\gamma = 1.4$. Also establish the velocity at $x' = 0.5$ when $t' = 0.4$.

*Solution:*

Since

$$u' = \frac{n' - m'}{\gamma - 1}$$

therefore

$$n'_A = u'_A(\gamma - 1) + m'_A$$

but

$$m'_A = m'_B$$

hence

$$n'_A = 0.2 \times 0.4 + 1.0$$

i.e.,

$$n'_A = 1.08$$

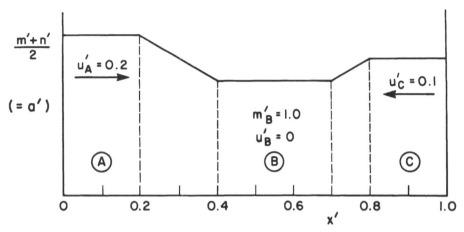

**Fig. 3–11** Initial conditions of wave-interaction example.

also, since $u'_B = 0$:
$$m'_B = n'_B = 1.0$$
and
$$m'_C = n'_C - (\gamma - 1)u'_C$$
but
$$n'_C = n'_B$$
hence
$$m'_C = 1.0 + 0.4 \times 0.1$$
i.e.,
$$m'_C = 1.04$$

It may be noted that the $m'$ and $n'$ values are incremented in quantities of either 0.04 or 0.08 for this particular problem, i.e., $m'_C = 1.04$, $n'_A = 1.08$.

Since more details of the wave events can be obtained when a large number of characteristics is drawn in a fine mesh rather than a small number in a coarse mesh, it would seem reasonable and convenient to choose $\pm 0.04$ as the increment between adjacent characteristics. It is *not* essential merely convenient, to choose a uniform incremental step between characteristics.

To determine the $x'$ value corresponding to $n' = 1.04$, since

$$\frac{m'_A + n'_A}{2} = \frac{2 \cdot 8}{2} = 1.04$$

and

$$\frac{m'_B + n'_B}{2} = \frac{2}{2} = 1.00$$

and $(m' + n')/2$ is linearly distributed between $x' = 0.2$; $x' = 0.4$, then for

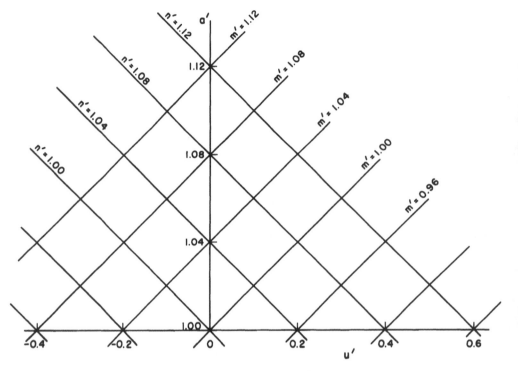

**Fig. 3–12** State diagram for wave-interaction example.

$n' = 1.04$,

$$\frac{m' + n'}{2} = \frac{2.04}{2} = 1.02$$

that, by linear interpolation, corresponds to $x' = 0.3$, i.e., $n' = 1.04$ is locate
at $x' = 0.3$ for the initial conditions of the problem.

The required portion of the $u' \sim a'$ diagram can now be constructed a
shown in Fig. 3.12.

The $x' \sim t'$ diagram can only be drawn after first evaluating the slopes 
the characteristics on the $x' \sim t'$ plane as shown in the following table:

Slope of an $n' = 1.00$ crossing an $m' = 1.00$ space

$$= \frac{1}{u' + a'} = \frac{1}{0 + 1.0} = 1.0$$

Slope of an $n' = 1.04$ crossing an $m' = 1.00$ space

$$= \frac{1}{u' + a'} = \frac{1}{0.1 + 1.02} = \frac{1}{1.12}$$

Slope of an $n' = 1.08$ crossing an $m' = 1.00$ space

$$= \frac{1}{u' + a'} = \frac{1}{0.2 + 1.04} = \frac{1}{1.24}$$

Slope of an $m' = 1.00$ crossing an $n' = 1.00$ space

$$= \frac{1}{u' - a'} = \frac{1}{0 - 1.0} = -1.0$$

Slope of an $m' = 1.04$ crossing an $n' = 1.00$ space

$$= \frac{1}{u' - a'} = \frac{1}{-0.1 - 1.02} = -\frac{1}{1.12}$$

Average slope of an $n' = 1.00$ between an $m' = 1.00$ and $m' = 1.04$

$$\simeq \frac{1}{\bar{u}' + \bar{a}'} = \frac{1}{-0.05 + 1.01} = \frac{1}{0.96}$$

Average slope of an $n' = 1.04$ between an $m' = 1.00$ and $m' = 1.04$

$$\simeq \frac{1}{\bar{u}' + \bar{a}'} = \frac{1}{0.05 + 1.03} = \frac{1}{1.08}$$

Average slope of an $n' = 1.08$ between an $m' = 1.00$ and $m' = 1.04$

$$\simeq \frac{1}{\bar{u}' + \bar{a}'} = \frac{1}{0.15 + 1.05} = \frac{1}{1.20}$$

Average slope of an $m' = 1.00$ between an $n' = 1.00$ and $n' = 1.04$

$$\simeq \frac{1}{\bar{u}' - \bar{a}'} = \frac{1}{0.05 - 1.01} = -\frac{1}{0.96}$$

Average slope of an $m' = 1.04$ between an $n' = 1.00$ and $n' = 1.04$

$$\simeq \frac{1}{\bar{u}' - \bar{a}'} = \frac{1}{-0.05 - 1.03} = -\frac{1}{1.08}$$

Average slope of an $m' = 1.00$ between an $n' = 1.04$ and $n' = 1.08$

$$\simeq \frac{1}{\bar{u}' - \bar{a}'} = \frac{1}{0.15 - 1.03} = -\frac{1}{0.88}$$

Average slope of an $m' = 1.04$ between an $n' = 1.04$ and $n' = 1.08$

$$\simeq \frac{1}{\bar{u}' - \bar{a}'} = \frac{1}{0.05 - 1.05} = -\frac{1}{1.0}$$

Slope of an $n' = 1.00$ crossing an $m' = 1.04$ space

$$= \frac{1}{u' + a'} = \frac{1}{-0.1 + 1.02} + \frac{1}{0.92}$$

Slope of an $n' = 1.04$ crossing an $m' = 1.04$ space

$$= \frac{1}{u' + a'} = \frac{1}{0 + 1.04} = \frac{1}{1.04}$$

Slope of an $n' = 1.08$ crossing an $m' = 1.04$ space

$$= \frac{1}{u' + a'} = \frac{1}{0.1 + 1.06} = \frac{1}{1.16}$$

Slope of an $m' = 1.00$ crossing an $n' = 1.08$ space

$$= \frac{1}{u' - a'} = \frac{1}{0.2 - 1.04} = -\frac{1}{0.84}$$

Slope of an $m' = 1.04$ crossing an $n' = 1.08$ space

$$= \frac{1}{u' - a'} = \frac{1}{0.1 - 1.06} = -\frac{1}{0.96}$$

The $x' \sim t'$ diagram, Fig. 3.13, can now be constructed. By reading from

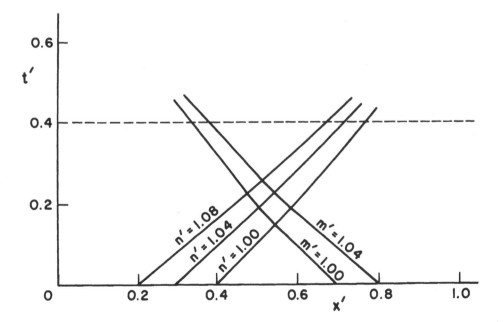

**Fig. 3–13** Wave diagram for wave-interaction example.

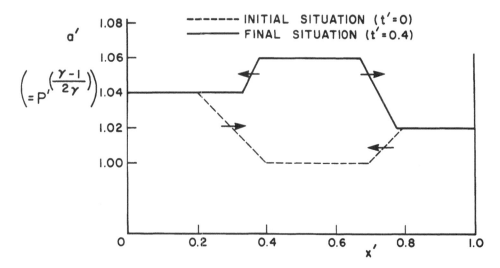

**Fig. 3–14** Comparison of initial and final conditions on the $x' \sim a'$ plane for the wave-interaction example.

data for $t' = 0.4$ on the $x' \sim t'$ plane an evaluation of $a'$ $(=P'^{(\gamma-1)/2\gamma})$ can be made and plotted versus $x'$; such a plot is presented in Fig. 3.14.

From the wave, or $x' \sim t'$, diagram when $t' = 0.4$, $n' = 1.08$, and $m' = 1.04$ at $x' = 0.5$. Hence

$$u' = \frac{n' - m'}{\gamma - 1} = \frac{1.08 - 1.04}{0.4}$$

therefore

$$u'_{(x'=0.5,\, t'=0.4)} = 0.1$$

Alternatively this result could have been obtained directly from the $u' \sim a'$ diagram for the values of $n'$ and $m'$ read from the $x' \sim t'$ diagram. The foregoing evaluation of $u'$ for $x' = 0.5$ and $t' = 0.4$, plus Fig. 3.14, constitute the required solution to the problem.

## Problems

3.1 Show, from first principles, using the same definitions of $f$ and $q$ as in Chapter 3, that for one-dimensional time-dependent flow with friction and heat transfer in a duct of nonuniform cross-secional area the equations of conservation of (a) mass, (b) momentum, and (c) energy are:

a. $\dfrac{\partial \rho}{\partial t} + \dfrac{\partial}{\partial x}(\rho u) + \dfrac{\rho u}{A}\dfrac{dA}{dx} = 0$

b. $\dfrac{\partial}{\partial t}(\rho u) + \dfrac{\partial}{\partial x}(\rho u^2 + P) + \dfrac{\rho u^2}{A}\dfrac{dA}{dx} = -f\rho$

c. $\dfrac{\partial}{\partial t}\left\{\rho\left[h+\dfrac{u^2}{2}-\dfrac{P}{\rho}\right]\right\}+\dfrac{\partial}{\partial x}\left\{\rho u\left[h+\dfrac{u^2}{2}\right]\right\}+\dfrac{1}{A}\left\{\rho u\left[h+\dfrac{u^2}{2}\right]\right\}\dfrac{dA}{dx}=q\rho$

3.2 Show that the equations of conservation of (a) mass, (b) momentum, and (c) energy for one-dimensional nonsteady compressible flow with friction and heat addition in a duct of nonuniform cross-sectional area can be written:

a. $\dfrac{\partial\rho}{\partial t}+u\dfrac{\partial\rho}{\partial x}+\rho\dfrac{\partial u}{\partial x}=-\dfrac{\rho u}{A}\dfrac{dA}{dx}$

b. $\dfrac{\partial u}{\partial t}+u\dfrac{\partial u}{\partial x}+\dfrac{1}{\rho}\dfrac{\partial P}{\partial x}=-f$

c. $\dfrac{\partial h}{\partial t}+u\dfrac{\partial h}{\partial x}-\dfrac{1}{\rho}\left\{\dfrac{\partial P}{\partial t}+u\dfrac{\partial P}{\partial x}\right\}=q+uf$

3.3 Starting with the First Law of Thermodynamics and taking account of the Second Law, derive the relationship:

$$T\,ds = dh - dP/\rho$$

3.4 Derive the following thermodynamically based relationships:

a. $\dfrac{\partial\rho}{\rho}=\left(\dfrac{2}{\gamma-1}\right)\dfrac{da}{a}-\dfrac{ds}{R}$

b. $\dfrac{\partial P}{P}=\left(\dfrac{2\gamma}{\gamma-1}\right)\dfrac{da}{a}-\dfrac{ds}{R}$

c. $dh-\dfrac{dP}{\rho}=\dfrac{1}{\gamma}\dfrac{a^2}{R}\,ds$

3.5 Where the differential operator $S$ (the substantial, or convective, derivative) is defined thus:

$$S\equiv\dfrac{\partial}{\partial t}+u\dfrac{\partial}{\partial x}$$

derive the equations of conservation of (a) momentum, (b) energy, and (c) mass in the following forms applicable to nonsteady, one-dimensional flow in a duct of nonuniform cross section:

a. $S\left(\dfrac{\gamma-1}{2}u\right)+a\dfrac{\partial a}{\partial x}=\dfrac{a^2}{2c_p}\dfrac{\partial s}{\partial x}-\left(\dfrac{\gamma-1}{2}\right)f$

b. $Ss=\dfrac{\gamma R}{a^2}(q+uf)$

c. $Sa+a\dfrac{\partial}{\partial x}\left(\dfrac{\gamma-1}{2}u\right)=\left(\dfrac{\gamma-1}{2}\right)\left\{\dfrac{\gamma}{a}(q+uf)-\dfrac{au}{A}\dfrac{dA}{dx}\right\}$

3.6 Illustrate, on the $x \sim t$ plane, the physical meaning of the differential operator $M$ operating on $\phi$ where:

$$M \equiv \frac{\partial}{\partial t} + (u - a)\frac{\partial}{\partial x}, \qquad \phi = \phi(x, t)$$

3.7 Show that the nondimensional differential operator $N'$ can be described in terms of the dimensional form $N$ by:

$$N' = (l/a_{\text{REF}})N$$

where:

$$l = \text{a reference length,}$$

$$a_{\text{REF}} = \text{a reference acoustic velocity.}$$

Show also that this result is consistent with the definitions of $a'$, $dA'/dx'$, $f'$, $n'$, $q'$, $s'$, $u'$, and $x'$ given in Chapter 3.

3.8 A stepped tube is arranged as shown in Fig. 3.15. The tube is provided with a puncturable diaphragm at the large diameter end. The ratio of the cross-sectional area of the large to the small diameter sections is very great (say several orders of magnitude). The stepped tube contains, initially, compressed air ($\gamma = 1.4$), at rest, at a pressure of 1.315 atmospheres and a temperature of 288 K. The pressure of the fluid surrounding the tube is 1.0 atmosphere. Evaluate, by means of the method-of-characteristics, the lowest pressure achieved, and the time taken to attain this condition at the blanked-off end of the small tube following instantaneous rupture of the diaphragm. Assume that the flow processes are isentropic and assume also that, due to the very large ratio of the area of the large diameter to the small diameter portion of the tube, flow into or out of the small diameter region has no influence on the flow field in the large diameter section.
*Note:* The open (diaphragm) end and closed (blanked-off) end boundary condition treatments of Chapter 2 are applicable to this problem.

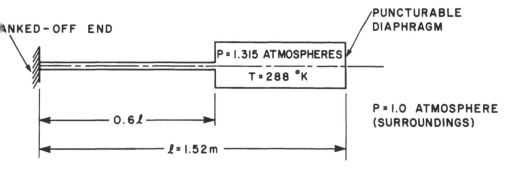

**Fig. 3–15** Diagram for problem 3.8.

# 4

## BOUNDARY CONDITIONS

Boundary conditions have already been mentioned briefly in the three previous chapters. Very basic boundary conditions were specified for simple isentropic nonsteady flow problems in Chapter 2. In Chapter 3 the need was emphasized for generally applicable rules to make method-of-characteristics computations feasible in the region of the flow-field boundaries. The boundary conditions that must be specified are normally those defining the:

i. flow field at $t' = t'_{\text{INITIAL}}$ as a function of $x'$ over the full range of $x'$, usually given by $0 \leq x' \leq 1$,

ii. flow conditions at each end of the duct, or tube, containing the nonsteady flow field. The duct ends are usually associated with $x' = 0$ and $x' = 1$; the boundary conditions at these stations, or their equivalents, must each be identified as a function of time $t'$.

Thus the boundaries along which sufficient conditions must be specified in order to predict the flow field can usually be represented by the shaded boundary of Fig. 4.1. It should be emphasized that Fig. 4.1 is an attempt to depict the most likely circumstances only. It is quite possible if, for example, a telescopic duct

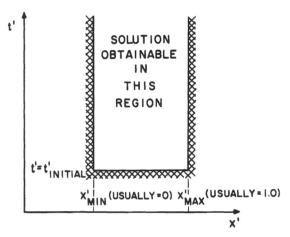

Fig. 4–1 Typical boundaries of a solution domain.

is employed for $x'_{MIN}$ and/or $x'_{MAX}$ to be functions of time. Furthermore, it is possible, but perhaps unlikely, that the $t' = t'_{INITIAL}$ boundary may be replaced by a suitable path through the flow along which the initial conditions are specified.

The boundary conditions specified in this chapter for duct ends are, perhaps, the simplest ones capable of representing, with reasonable accuracy, conditions applicable to many nonsteady flow situations. Most appear to have been originated by Jenny (1950) for use in reciprocating internal-combustion engine exhaust-system flow analysis.

The purpose of boundary conditions is, apart from specification of the initial conditions within the duct or pipe (i.e., the initial spatial boundary condition), to provide data for the left and right-hand sides, or boundaries, of the flow field which, due to the presence of a boundary, are not forthcoming from the method-of-characteristics. Figure 4.2 is an attempt to illustrate the situations that can arise at open, or partially open, ends. As can be seen from the upper part of Fig. 4.2, for subsonic flow from left to right, boundary node $A$ lies at an inlet and it is clear that the $n'$ and $s'$ characteristics for node $A$ do not exist. The information they would have contributed must be supplied by the boundary condition equations. For the sake of comparison, node $B$ is a nonboundary point and here all three characteristics exist; hence boundary equations are not needed for this node and all others that do not lie on the flow-field boundaries. Node $C$ represents an outlet; for this case the boundary condition must supply information corresponding to that which would have been provided by the $m'$ characteristic had it existed.

The lower part of Fig. 4.2 illustrates the corresponding situation for flow from right to left. It is apparent from both portions of Fig. 4.2 that more is demanded of boundary condition equations for inlets than for outlets. This is because inlet boundary conditions have to provide the equivalent of information corresponding to two characteristics compared with the equivalent of the information relating to only one characteristic for an outlet. Boundary condition equations for inlets and outlets are also called upon to identify the type of flow occurring and to classify it as subcritical or supercritical. This is necessary since these two classes of flow are handled differently. The simplest form of boundary condition is that for a closed end.

There are inherently two versions of each boundary condition when written in terms of the characteristic quantities $m'$, $n'$, and $s'$. One version is applicable at a right-hand boundary, the other at a left-hand boundary. For the sake of simplicity only a single version of each boundary condition is presented; it is expressed in terms of the known, prescribed, boundary conditions and the $a'$, $u'$, and $s'$ variables of the nonsteady flow field.

All the open end boundary conditions employed on the $x' = 0$, and $x' = 1.0$ boundaries (i.e., at duct ends) are based on the assumption of quasi-steady flow conditions and were, therefore, obtained assuming that, at any instant, conventional steady-flow analyses can be employed to describe the instantaneous flow conditions into, or out of, the duct in which the unsteady flow occurs.

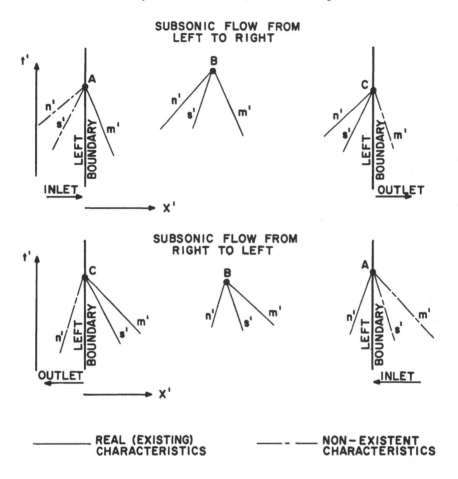

**Fig. 4–2** Characteristics for nodal points located on the left- and right-hand boundaries of the solution domain.

### 4.1   Initial Spatial Boundary Condition

To commence to obtain a solution it is first necessary to specify the initial conditions in the nonsteady flow field. This, of course, implies a physical knowledge of the problem that is equivalent to specifying the distribution of the three characteristic quantities

$$m', n', \text{ and } s'$$

each as a function of $x'$ in the range; $0 \leq x' \leq 1.0$ for $t' = t'_{\text{INITIAL}}$. Often $t'_{\text{INITIAL}} = 0$.

Frequently the initial conditions are known from the physical description of the problem, for example, the fluid may be at rest and of uniform entropy. Thus, for particular cases in which the fluid is at rest $u' = 0$ and hence from Eq. (3.69) (Chapter 3), $m' = n'$.

However, in many problems in which the unsteady process is cyclic the fluid will not be at rest initially. Thus, without knowing the solution in advance, it is impossible to specify correctly the required information. In such cases it is usual to make a simplied, physically reasonable guess at the initial conditions and to perform successive calculations by the method-of-characteristics until cyclic conditions have been established. Cyclic operation is identified by the requirement that the distributions of the characteristic quantities, as functions of $x'$, are the same at the end of a cycle as they were at its beginning. It is only really practical to carry out an iterative calculation of this kind, using the method-of-characteristics, when a computer is available.

The foregoing rather general remarks suffice to explain the setting up of boundary conditions relating to $t' = t'_{INITIAL}$, $0 \le x' \le 1.0$ (or corresponding relevant values of $x'$ for a particular problem).

The boundary conditions for $x' = 0$ and $x' = 1.0$ for all $t'$ must now be considered.

## 4.2 Closed Ends

For this condition:

$$u' = 0 \tag{4.1}$$

As a consequence of this it can be seen that, from Eqs. (3.62) and (3.69) of Chapter 3, at a closed end:

$$m' = n' \tag{4.2}$$

$$\beta'_s = 0 \tag{4.3}$$

The entropy boundary condition depends upon the relationship given by Eqs. (3.65) and (3.58). Hence when friction and heat transfer are absent entropy is constant adjacent to a closed end.

In some circumstances a closed end is replaced by a moving piston. In such cases the zero on the right-hand side of Eq. (4.1) is replaced by the normalized piston velocity $u'_{PISTON}$. This results in corresponding changes in Eqs. (4.2) and (4.3).

## 4.3 Outlets

Assumptions implicit in the setting up of boundary conditions for outlets are as follows:

  i. wall friction and heat transfer are absent in an outlet region,
 ii. flow out of a fully or partly open outlet is assumed to be isentropic (up to the throat of the exit),

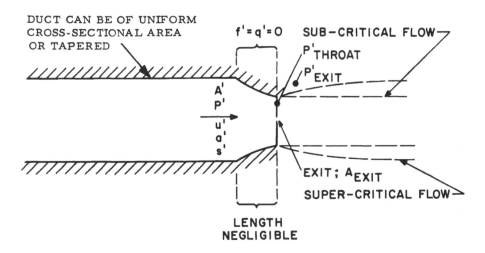

DUCT CAN BE OF UNIFORM
CROSS-SECTIONAL AREA
OR TAPERED

$f' = q' = 0$      SUB-CRITICAL FLOW

$P'_{\text{THROAT}}$

$P'_{\text{EXIT}}$

$A'$
$P'$
$u'$
$a'$
$s'$

EXIT; $A_{\text{EXIT}}$

SUPER-CRITICAL FLOW

LENGTH
NEGLIGIBLE

**Fig. 4–3** Diagrammatic illustration of a partly open outlet.

iii. vena-contracta effects are absent,

iv. the axial distance occupied by an outlet and its associated flow field is small compared with the length of the duct,

v. flow is one-dimensional in the vicinity of an outlet.

An outlet is represented pictorially in Fig. 4.3. The nozzle shown in the exit region of the duct should be interpreted rather loosely as literally a nozzle of fixed area ratio as indicated or as a nozzle the exit area of which varies as a function of time. This latter situation applies when the nozzle represents a valve, for example an inlet valve of a reciprocating, four-stroke, internal-combustion engine. For cases of this type:

$$0 \le \Phi \le 1$$

where

$$\Phi \equiv \frac{C_f A_{\text{EXIT}}}{A} \tag{4.4}$$

Area $A$ in Eq. (4.4) is the duct cross-sectional area adjacent to the outlet. The coefficient $C_f$ of (4.4) is a prescribed function of $A_{\text{EXIT}}/A$; the purpose of which is·to take into account boundary layer effects, etc., in the exit and, partially at least, vena-contracta effects. To take vena-contracta effects into account rigorously $C_f$ would additionally need to be a function of the pressure ratio of the discharge. To introduce such a comprehensive coefficient would complicate the boundary condition equations although, presumably, it could be incorporated if the need were sufficiently pressing, which is unlikely to be the case for most problems. Hence for the cases to be considered assumption (iii) is for all practical purposes true for the outflow cases that can be handled.

### 4.3.1 Specification of boundary data for outflow

In order that the outflow conditions can be specified it is necessary to provide two pieces of information:

i. the normalized static pressure at the exit, $P'_{EXIT}$, as a function of $t'$ or the normalized stagnation pressure, $P'_{oEXIT}$, also as a function of time $t'$.

   Note: The static pressure is most likely to be the known condition in an outlet; alternatively the stagnation pressure may be given, however, but this is only realistic when some provision is made to recover flow kinetic energy downstream of the outlet.

ii. the ratio $\Phi$ $(0 \leq \Phi \leq 1.0)$ as a function of time $t'$.

It is important to note that the purpose of the following boundary conditions is to relate the nonsteady flow condition within the duct, pipe, etc., to the prescribed conditions controlling flow in the open duct end and *not* to relate the flow conditions in the duct to those in the exit throat itself.

### 4.3.2 Subcritical outflow

For compressible flow, the static to stagnation pressure ratio in the duct is given by

$$\frac{P'_o}{P'} = \left\{ 1 + \frac{\gamma - 1}{2} \left( \frac{u'}{a'} \right)^2 \right\}^{\gamma/(\gamma - 1)} \tag{4.5}$$

this is a normal steady-flow isentropic relationship.

The condition—throat Mach number $\leq 1$ for subcritical operation is given by

$$\frac{P'_o}{P'_{EXIT}} \leq \left( \frac{\gamma - 1}{2} \right)^{\gamma/(\gamma - 1)} \tag{4.6}$$

(Note that $P'_{THROAT} = P'_{EXIT}$ when the flow is subcritical and that the subscripts "exit" and "throat" refer to the geometric throat or exit or, if there is a vena-contracta, the vena-contracta location.)

Hence from Eqs. (4.5) and (4.6), and also Eq. (3.72) of Chapter 3, the condition for subcritical outflow becomes:

$$P'_{EXIT} \geq (a')^{2\gamma/(\gamma - 1)} e^{-s'} \left\{ \frac{2}{\gamma + 1} + \frac{\gamma - 1}{\gamma + 1} \left( \frac{u'}{a'} \right)^2 \right\}^{\gamma/(\gamma - 1)} \tag{4.7}$$

where $u'$, $s'$, and $a'$ refer to conditions in the duct, or pipe, in flow approaching the exit but which has not entered the contraction leading to the exit.

*Fully open outlets ($P'_{\text{EXIT}}$ prescribed)*

When the outlet end of the duct or pipe is fully uncovered ($\Phi = 1$) and $P'_{\text{EXIT}} = P'$ hence:

$$P'_{\text{EXIT}} = (a')^{2\gamma/(\gamma-1)}e^{-s'} \qquad (4.8)$$

From inspection of Eq. (3.68) of Chapter 3 it can be seen that either $m'$ or $n'$ can be established provided the other is known. Thus for an outlet at the right-hand end of a pipe or duct $n'$ and $s'$ are both known from the flow field within the pipe or duct and hence $m'$ can, therefore, be found from (4.8) in conjunction with Eq. (3.68).

Conversely, for an outlet at the left-hand end, $m'$ and $s'$ are known and $n'$ is found from (4.8) and Eq. (3.68).

*Partly open outlets ($P'_{\text{EXIT}}$ prescribed)*

Application of the steady-flow continuity equation to the outlet gives

$$u' = \Phi\,\frac{P'_{\text{EXIT}}}{P'}\,\frac{T'}{T_{\text{EXIT}}} \cdot u'_{\text{EXIT}} = \Phi\left(\frac{P'_{\text{EXIT}}}{P'}\right)^{1/\gamma} u'_{\text{EXIT}} \qquad (4.9)$$

Here $u'$ and $u'_{\text{EXIT}}$ are positive or negative in accordance with the flow direction.

Invoking Eq. (4.5) and applying it to the flow in the port gives, after rearrangement:

$$\left\{\frac{u'_{\text{EXIT}}}{a'_{\text{EXIT}}}\right\}^2 = \frac{2}{\gamma-1}\left\{\left(\frac{P'_0}{P'_{\text{EXIT}}}\right)^{(\gamma-1)/\gamma} - 1\right\} \qquad (4.10)$$

A further application of (4.5) allows $P'_0$ in (4.10) to be replaced in terms of $u'/a'$ and $P'$ thus

$$\left(\frac{u'_{\text{EXIT}}}{a'_{\text{EXIT}}}\right)^2 = \frac{2}{\gamma-1}\left[\left(\frac{P'}{P'_{\text{EXIT}}}\right)^{(\gamma-1)/\gamma}\left\{1 + \frac{\gamma-1}{2}\left(\frac{u'}{a'}\right)^2\right\} - 1\right] \qquad (4.11)$$

*Note:* This is possible because the stagnation pressure is the same in the exit port as that in the pipe, or duct, in the flow approaching the exit (isentropic flow).

Eliminating $u'_{\text{EXIT}}$ from (4.11) by substitution from (4.9) gives:

$$\left(\frac{u'}{a'_{\text{EXIT}}}\right)^2 = \frac{\dfrac{2}{\gamma-1}\left\{\left(\dfrac{P'}{P'_{\text{EXIT}}}\right)^{(\gamma-1)/\gamma} - 1\right\}}{\dfrac{1}{\Phi^2}\left(\dfrac{P'}{P'_{\text{EXIT}}}\right)^{2/\gamma} - 1} \qquad (4.12)$$

Invoking the isentropic relationship Eq. (3.72) and substituting in (4.12) for

$a'_{EXIT}$ in terms of $a'$ gives:

$$\left(\frac{u'}{a'}\right)^2 = \frac{\dfrac{2}{\gamma-1}\left\{1 - \left(\dfrac{P'}{P'_{EXIT}}\right)^{-(\gamma-1)/\gamma}\right\}}{\dfrac{1}{\Phi^2}\left(\dfrac{P'}{P'_{EXIT}}\right)^{2/\gamma} - 1}$$

or by replacing $P'$ in terms of $a'$ and $s'$ from Eq. (3.72):

$$\left(\frac{u'}{a'}\right)^2 = \frac{\dfrac{2}{\gamma-1}\left\{1 - (a')^{-2}\left(\dfrac{e^{-s'}}{P'_{EXIT}}\right)^{-(\gamma-1)/\gamma}\right\}}{\dfrac{a'^{4/(\gamma-1)}}{\Phi^2}\left(\dfrac{e^{-s'}}{P'_{EXIT}}\right)^{2/\gamma} - 1} \tag{4.13}$$

Again, for prescribed $P'_{EXIT}$ and $\Phi$, the only unknown at a right-hand outlet is $m'$, both $n'$ and $s'$ being known. For a left-hand outlet the unknown is $n'$, $m'$ and $s'$ both being known.

It is worth noting that when $\Phi = 0$ Eq. (4.13) is consistent with the closed-end boundary condition given by Eq. (4.1). When $\Phi = 1$ Eq. (4.13) is consistent with the fully open boundary condition defined in Eq. (4.8).

### Prescribed outlet port stagnation pressure (fully or partly open)

When an outlet stagnation pressure $P'_{oEXIT}$ is prescribed (instead of the static pressure $P'_{EXIT}$) substitution in Eq. (4.5) for $P'$ from Eq. (3.72) gives the required relationship between the known boundary condition and flow in the duct or pipe, etc., thus

$$P'_{oEXIT} = (a')^{2\gamma/(\gamma-1)}e^{-s'}\left\{1 + \frac{\gamma-1}{2}\left(\frac{u'}{a'}\right)^2\right\}^{\gamma/(\gamma-1)} \tag{4.14}$$

This result applies to both fully open and partly open duct ends. The absence of dependence upon $\Phi$ is noteworthy. It arises from the assumption that the outflow process is isentropic for all $\Phi$, hence $P'_{oEXIT} = P'_o$.

### 4.3.3 Supercritical outflow

The condition for supercritical operation is obtained from (4.7); thus outflow is supercritical when:

$$P'_{EXIT} < (a')^{2\gamma/(\gamma-1)}e^{-s'}\left\{\frac{2}{\gamma+1} + \frac{\gamma-1}{\gamma+1}\left(\frac{u'}{a'}\right)^2\right\}^{\gamma/(\gamma-1)} \tag{4.15}$$

### Fully open outlets

An expansion wave propagating upstream cannot accelerate a flow in a duct sufficiently to achieve a Mach number greater than unity. The reason is that when the local flow attains a velocity equal and opposite to the local acoustic velocity there can be no further propagation of additional expansion waves in an upstream direction since the wave-propagation velocity in the upstream direction is given by

$$a' - |u'| = 0$$

Hence for such circumstances at a duct open end:

$$\frac{u'}{a'} = \pm 1 \qquad (4.16)$$

provided the condition represented by Eq. (4.15) also applies. The positive or negative signs correspond to the flow direction.

However, a supersonic flow may exist in the duct initially or it can be generated by compression waves propagating toward the open end. In either case when supersonic flow prevails in the duct or pipe at the exit plane, the external conditions have no effect on the flow provided the static pressure in the region into which the discharge takes place does not exceed that achieved when a normal shock wave stands in the duct exit. This condition is quantified using Eq. (2.10) a well-known static pressure ratio relationship for a normal shock wave in steady flow:

$$\frac{P'}{P'_{\text{EXIT}}} > \frac{2\gamma\left(\dfrac{u'}{a'}\right)^2 - (\gamma - 1)}{\gamma + 1}$$

or

$$P'_{\text{EXIT}} < \frac{(\gamma + 1)(a')^{2\gamma/(\gamma - 1)}e^{-s'}}{2\gamma\left(\dfrac{u'}{a'}\right)^2 - (\gamma - 1)} \qquad (4.17)$$

where $a'$, $u'$, and $s'$ refer to the supersonic flow adjacent to the outlet.

A supersonic outflow at the boundary of a nonsteady, compressible flow field does not seem to be a very common situation in most practical problems although it is one that can occur. A simple configuration that can produce such a flow condition is illustrated in Fig. 4.4; the diagram shows a convergent-divergent nozzle supplied from a very large pressure vessel. The exit of the nozzle is provided with a frangible diaphragm. Destruction of the diaphragm results in flow initiation by means of expansion waves. Simplifying the flow description, the expansion waves are reflected from the upstream end of the nozzle as compression waves. It is apparent, from knowledge of steady flow, that ultimately a steady supersonic flow will prevail at the nozzle exit, after the transient flow regime has subsided, provided conditions (4.15) and (4.17) are both satisfied. Condition (4.15) is recognizable as a statement that the ratio of the stagnation pressure at the nozzle exit (and hence, for steady flow, everywhere else within the nozzle) to that of the surroundings must exceed the critical pressure ratio $((\gamma + 1)/2)^{\gamma/((\gamma - 1))}$.

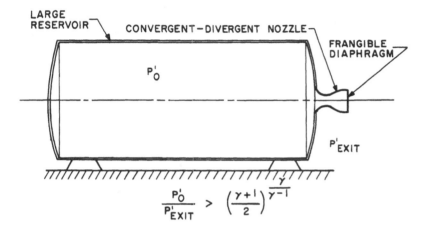

**Fig. 4–4** System capable of producing supersonic outflow following an initiating expansion wave.

*Partly open outlets*

Application of the steady-flow continuity equation to the flow approaching the exit, and that passing through the exit, together with the condition that $|u'_{\text{THROAT}}|/a'_{\text{THROAT}} = 1$ gives

$$\frac{|u'|}{a'_{\text{THROAT}}} = \Phi\left(\frac{P'_{\text{THROAT}}}{P'}\right)^{1/\gamma} \tag{4.18}$$

Invoking the isentropic relationship, Eq. (3.72), between $a'$ and $P'$ and substituting for $a'_{\text{THROAT}}$ in (4.18) yields

$$\frac{|u'|}{a'} = \Phi\left(\frac{P'_{\text{THROAT}}}{P'}\right)^{(\gamma+1)/2\gamma} \tag{4.19}$$

From Eq. (4.5) and the condition for choking flow,

$$P'_o/P'_{\text{THROAT}} = \left(\frac{\gamma+1}{2}\right)^{\gamma/(\gamma-1)}$$

the following result is obtainable:

$$P'_{\text{THROAT}} = P'\left\{\frac{2}{\gamma+1} + \frac{\gamma-1}{\gamma+1}\left(\frac{u'}{a'}\right)^2\right\}^{\gamma/(\gamma-1)} \tag{4.20}$$

Substituting for $P'_{\text{THROAT}}$ in (4.19) from (4.20):

$$\frac{|u'|}{a'} = \Phi\left\{\frac{2}{\gamma+1} + \frac{\gamma-1}{\gamma+1}\left(\frac{u'}{a'}\right)^2\right\}^{(\gamma+1)/2(\gamma-1)} \tag{4.21}$$

Here, for an opening at the right-hand end of a duct, $n'$ and $s'$ are known and only $m'$ is to be established from (4.21). When the opening is at the left $m'$ and $s'$ will be known and $n'$ has to be found. Note that (4.21) is independent of $P'_{EXIT}$.

It can be shown that when $\Phi = 0$ Eq. (4.21) is consistent with the closed-end boundary condition, Eq. (4.1), and when $\Phi = 1$ it is consistent with the sonic flow condition defined in Eq. (4.16). Equation (4.16) is in turn consistent, for $\Phi = 1$, with the condition that $|u'_{THROAT}|/a'_{THROAT} = 1$, which was incorporated in the derivation of Eq. (4.21).

*Prescribed outlet port stagnation pressure (fully or partly open)*

Since the outflow is assumed to be isentropic, even when supersonic, the same result is obtained as for the subcritical case. Hence Eq. (4.14) is also applicable to supercritical operation.

## 4.4  Inlets

The assumptions relating to the inlet boundary equations are as follows:

  i.  wall friction and heat transfer are absent in the inlet region,
  ii.  pressure recovery downstream of a partly open inlet is ignored at subcritical pressure ratios,
  iii.  flow into a fully open inlet is assumed to be isentropic,
  iv.  flow from the port, or opening, up to the throat of the inlet restriction is assumed to be isentropic for a partly open inlet,
  v.  vena-contracta effects are ignored,
  vi.  the axial distance occupied by the inlet and its associated flow field is small compared with the length of the duct containing the unsteady flow field,
  vii.  flow is one-dimensional in the vicinity of the inlet.

*Note:* Assumption (ii) is slightly pessimistic but it does permit simplification of the equations; it is in accordance with Jenny's assumptions (Jenny, 1950). Assumption (vi) is not always accurate as the entering flow may penetrate for an appreciable distance into a duct, or pipe, before becoming one-dimensional. The assumption, (v), that vena-contracta effects are ignored is, in the rigorous sense, true although such effects can be included partially in the flow coefficient, $C_f$, of Eq. (4.22). By analogy with the situation for outlets, $C_f$ can be a prescribed function of $A_{INLET}/A$ but there is no provision to make $C_f$ also a function of pressure ratio as that is necessary if vena-contracta effects are to be accounted for completely. The effective open area as a fraction of the inlet plane cross-sectional area is given by $\Phi$:

$$\Phi \equiv C_f A_{INLET}/A \qquad (4.22)$$

An inlet region is represented diagrammatically in Fig. 4.5.

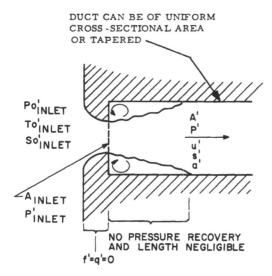

DUCT CAN BE OF UNIFORM
CROSS-SECTIONAL AREA
OR TAPERED

$Po'_{INLET}$
$To'_{INLET}$
$So'_{INLET}$

A'
P'
u'
s'
a'

$A_{INLET}$
$P'_{INLET}$

NO PRESSURE RECOVERY
AND LENGTH NEGLIGIBLE
$f'=q'=0$

$S'_{INLET}=So'_{INLET}$

**Fig. 4–5** Diagammatic illustration of a partly open inlet.

### 4.4.1 Specification of boundary data for inflow

For inflow three independent parameters must be specified. The most usual and convenient of these are:

   i. the normalized stagnation pressure $P'_{oINLET}$ as a function of $t'$,
   ii. the normalized stagnation temperature $T'_{oINLET}$ as a function of $t'$,
   iii. the ratio $\Phi$ also as a function of $t'$.

The entropy of the inflow is also required but this can be evaluated from $P'_{oINLET}$ and $T'_{oINLET}$ from Eqs. (3.71) and (3.72), thus:

$$P'_{oINLET} = (T'_{oINLET})^{\gamma/(\gamma-1)} e^{-s'_{oINLET}}$$

and hence by rearrangement of the logarithmic form of this equation

$$S'_{oINLET} = \ln\left\{\frac{(T'_{oINLET})^{\gamma/(\gamma-1)}}{P'_{oINLET}}\right\} \tag{4.23}$$

In a manner corresponding to that for the outflow case, the purpose of the inflow boundary equations is to relate the prescribed conditions at the inlet port, or in the duct entry, with those prevailing within the duct, *not* with those in the throat formed by the partly open inlet.

### 4.4.2   Subcritical inflow

The condition for subcritical operation (i.e., inlet throat Mach number $<1$) is given by

$$\frac{P'_{\text{oINLET}}}{P'_{\text{INLET}}} < \left(\frac{\gamma+1}{2}\right)^{\gamma/(\gamma-1)} \tag{4.24}$$

However, since there is no pressure recovery [assumption (ii)] then

$$P' = P'_{\text{INLET}} \tag{4.25}$$

thus from (4.24), (4.25), and Eq. (3.72) of Chapter 3, the condition for subcritical inflow is given by

$$P'_{\text{oINLET}} < \left(\frac{\gamma+1}{2}\right)^{\gamma/(\gamma-1)} (a')^{2\gamma/(\gamma-1)} e^{-s'} \tag{4.26}$$

It is important to note that this inequality, while perfectly correct, is not, in general, so easy to use as its outflow counterpart. The known values are $P'_{\text{oINLET}}$ and either $n'$ or $m'$ depending upon whether the inlet is on the right or left-hand end, respectively, of the duct, exhaust pipe, etc. Since the equation relates to inflow, the entropy $s'$ is not known unless the inlet port is fully uncovered, in which case $s' = s'_{\text{oINLET}}$. For cases in which the end is only partly uncovered $s' > s'_{\text{oINLET}}$ due to irreversibilities downstream of the inlet throat.

### Fully open inlets

When an inlet is fully open the inflow process is, as previously stated, fully isentropic hence

$$s' = s'_{\text{oINLET}} \tag{4.27}$$

From application of Eq. (4.5) to inflow through a fully open inlet and also from substitution of $P'$ for $P'_{\text{INLET}}$ [Eq. (4.25)]:

$$\frac{P'_{\text{oINLET}}}{P'} = \left\{1 + \frac{\gamma-1}{2}\left(\frac{u'}{a'}\right)^2\right\}^{\gamma/(\gamma-1)}$$

$$= \left\{1 + \frac{\gamma-1}{2}\left(\frac{u'}{a'_{\text{oINLET}}}\right)^2 \left(\frac{a'_{\text{oINLET}}}{a'}\right)^2\right\}^{\gamma/(\gamma-1)}$$

substituting from Eq. (3.72) to replace $a'_{\text{oINLET}}/a'$ in terms of pressure ratio:

$$\frac{P'_{\text{oINLET}}}{P'} = \left\{1 + \frac{\gamma-1}{2}\left(\frac{u'}{a'_{\text{oINLET}}}\right)^2 \left(\frac{P'_{\text{oINLET}}}{P'}\right)^{\gamma/(\gamma-1)}\right\}^{\gamma/(\gamma-1)}$$

Rearranging, simplifying, and substituting for $P'$ from Eq. (3.72)

$$P'_{\text{oINLET}} = \frac{(a')^{2\gamma/(\gamma-1)} e^{-s'}}{\left\{1 - \frac{\gamma-1}{2}\left(\frac{u'}{a'_{\text{oINLET}}}\right)^2\right\}^{\gamma/(\gamma-1)}} \tag{4.28}$$

$s'$ is obtained from Eq. (4.27). Equation (4.28) permits either $n'$ or $m'$, as appropriate, to be established.

*Partly open inlets*

The entropy in the duct immediately downstream of the inlet is obtained by a procedure similar to that used for establishing the inlet stagnation entropy (Section 4.4.1):

$$s' = \ln\left\{\frac{(T')^{\gamma/(\gamma-1)}}{P'}\right\}$$

$$= \ln\left\{\frac{\left(\dfrac{T'}{T'_{\text{oINLET}}}\right)^{\gamma/(\gamma-1)}}{\dfrac{P'}{P'_{\text{oINLET}}}} \cdot \frac{(T'_{\text{oINLET}})^{\gamma/(\gamma-1)}}{P'_{\text{oINLET}}}\right\}$$

hence

$$s' = \ln\left\{\frac{\left(\dfrac{T'}{T'_{\text{oINLET}}}\right)^{\gamma/(\gamma-1)}}{\dfrac{P'}{P'_{\text{oINLET}}}}\right\} + s'_{\text{oINLET}} \tag{4.29}$$

Also it may be shown from the energy equation that

$$\frac{T'_{\text{oINLET}}}{T'} = \frac{1}{1 - \left(\dfrac{\gamma-1}{2}\right)\left(\dfrac{u'}{a'_{\text{oINLET}}}\right)^2} \tag{4.30}$$

Hence, by substituting in (4.29) for $T'/T'_{\text{oINLET}}$ from (4.30):

$$s' = \ln\left[\frac{P'_{\text{oINLET}}}{P'}\left\{1 - \frac{\gamma-1}{2}\left(\frac{u'}{a'_{\text{oINLET}}}\right)^2\right\}^{\gamma/(\gamma-1)}\right] + s'_{\text{oINLET}} \tag{4.31}$$

This is the required expression for $s'$.

Applying the steady-flow continuity equation to the inlet flow gives:

$$u' = \Phi\frac{P'_{\text{INLET}}}{P'} \cdot \frac{T'}{T'_{\text{INLET}}} \cdot u'_{\text{INLET}}$$

hence

$$\frac{u'}{a'_{\text{oINLET}}} = \Phi\frac{P'_{\text{INLET}}}{P'} \cdot \frac{T'}{T'_{\text{INLET}}} \cdot \frac{u'_{\text{INLET}}}{a'_{\text{INLET}}} \cdot \frac{a'_{\text{INLET}}}{a'_{\text{oINLET}}} \tag{4.32}$$

Substituting in Eq. (4.32) from Eqs. (4.5) and (4.30) allows the following result to be obtained, after some manipulation, bearing in mind (4.25) namely $P'_{\text{INLET}} = P'$:

$$\frac{\dfrac{u'}{a'_{\text{oINLET}}}}{1 - \dfrac{\gamma - 1}{2}\left(\dfrac{u'}{a'_{\text{oINLET}}}\right)^2} = \Phi \left(\frac{P'_{\text{oINLET}}}{P'}\right)^{(\gamma - 1)/2\gamma} \sqrt{\frac{2}{\gamma - 1}\left\{\left(\frac{P'_{\text{oINLET}}}{P'}\right)^{\gamma/(\gamma - 1)} - 1\right\}} \quad (4.33)$$

where the sign of the radical term must be consistent with that of $u'$. Equations (4.31) and (4.33) must be solved simultaneously as there are two unknowns, either $m'$ and $s'$ when the inlet port is on the right, or $n'$ and $s'$ when it is on the left. A possible, but by no means unique, iterative solution procedure is to guess a value of $m'$ or $n'$, which ever is unknown, and hence evaluate $u'$ from Eq. (3.69). Using this value of $u'$ evaluate $P'$ from Eq. (4.33). Substitution of the previously determined values of $P'$ and $u'$ in Eq. (4.31) allows $s'$ to be established. The accuracy of the original guess is checked with the aid of Eqs. (3.68) and (3.72) that can be combined to give:

$$P' = \left(\frac{m' + n'}{2}\right)^{2\gamma/(\gamma - 1)} e^{-s'}$$

If this value of $P'$ is not equal to the value obtained from (4.33) adjust the value of the unknown, $m'$ or $n'$, and repeat until convergence is achieved.

It can be shown that when $\Phi = 0$ Eq. (4.31) is redundant and Eq. (4.33) is consistent with the result $u' = 0$; this is in accordance with the boundary condition for a closed end [Eq. (4.1)]. When $\Phi = 1$, Eq. (4.31) is consistent with the expectation for isentropic inflow namely $s' = s'_{\text{oINLET}}$. Equation (4.33) is also consistent with the conditions for isentropic inflow when $\Phi = 1$.

### 4.4.3  Supercritical inflow

The condition for supercritical inflow is, from (4.26):

$$P'_{\text{oINLET}} \geq \left(\frac{\gamma + 1}{2}\right)^{\gamma/(\gamma - 1)} (a')^{2\gamma/(\gamma - 1)} e^{-s'} \quad (4.34)$$

Similar comments can be made here regarding the evaluation of (4.34) as were made following (4.26).

### Fully open inlets

As for the subcritical case:

$$s' = s'_{\text{oINLET}} \quad (4.35)$$

Also, since an inlet is a wholly convergent passage when fully open, $|u'|/a' = 1$, thus

$$P'_{\text{oINLET}} = \left(\frac{\gamma + 1}{2}\right)^{\gamma/(\gamma - 1)} (a')^{2\gamma/(\gamma - 1)} e^{-s'} \quad (4.36)$$

$s'$ is obtained from (4.35), either $m'$ or $n'$, as appropriate, from (4.36).

*Partly open inlets*

Here the assumption that $P'_{\text{INLET}} = P'$ is no longer made; with choked inflow in partly open ends $P'_{\text{INLET}} \geq P'$.

The entropy boundary condition is similar to that for the subcritical case, however, in the supercritical case it becomes necessary to eliminate $P'_{\text{oINLET}}/P'$ terms to enable the two simultaneous equations describing supercritical inflow to be set up in a soluble form.

Equation (4.32) can be rewritten:

$$\frac{u'}{a'_{\text{oINLET}}} = \Phi \frac{P'_{\text{oINLET}}}{P'} \cdot \frac{P'_{\text{INLET}}}{P'_{\text{oINLET}}} \cdot \frac{T'}{T'_{\text{INLET}}} \frac{u'_{\text{INLET}}}{a'_{\text{INLET}}} \cdot \frac{a'_{\text{INLET}}}{a'_{\text{oINLET}}}$$

which for choked flow reduces to

$$\frac{|u'|}{a'_{\text{oINLET}}} = \Phi \frac{P'_{\text{oINLET}}}{P'} \left\{ 1 - \frac{\gamma - 1}{2} \left( \frac{u'}{a'_{\text{oINLET}}} \right)^2 \right\} \left( \frac{2}{\gamma + 1} \right)^{(\gamma+1)/2(\gamma-1)} \tag{4.37}$$

Substitution for $P'_{\text{oINLET}}/P'$ in (4.31) from (4.37) gives:

$$s' = \ln \left[ \frac{1}{\Phi} \left( \frac{\gamma + 1}{2} \right)^{(\gamma+1)/2(\gamma-1)} \frac{|u'|}{a'_{\text{oINLET}}} \left\{ 1 - \frac{\gamma - 1}{2} \left( \frac{u'}{a'_{\text{oINLET}}} \right)^2 \right\}^{\gamma/(\gamma-1)} \right] + s'_{\text{oINLET}} \tag{4.38}$$

since for the choked condition

$$\frac{|u'_{\text{INLET}}|}{a'_{\text{INLET}}} = 1,$$

$$\frac{P'_{\text{oINLET}}}{P'_{\text{INLET}}} = \left( \frac{\gamma + 1}{2} \right)^{\gamma/(\gamma-1)} \tag{4.39}$$

Invoking the additional result (4.39) and bearing in mind that $P' \neq P'_{\text{INLET}}$ for choked flow, the following result is obtained in a similar manner to Eq. (4.33). Substitution is also made from Eq. (3.72):

$$\frac{\dfrac{|u'|}{a'_{\text{oINLET}}}}{1 - \dfrac{\gamma - 1}{2} \left( \dfrac{u'}{a'_{\text{oINLET}}} \right)^2} = \Phi \left( \frac{2}{\gamma + 1} \right)^{(\gamma+1)/2(\gamma-1)} \frac{P'_{\text{oINLET}}}{(a')^{2\gamma/(\gamma-1)} e^{-s'}} \tag{4.40}$$

Equations (4.38) and (4.40) must be solved simultaneously for the two unknowns $s'$ and either $m'$ or $n'$ depending upon whether the inlet is at the right- or left-hand end of the cell, or duct, respectively.

### 4.4.4  Inflow with pressure recovery

The pressure recovery which can, in principle, occur with partially open inlets was neglected in the treatment of subcritical inflow presented in Section 4.4.2.

Pressure recovery can be taken into account by introducing the conservation of momentum principle into the analysis of inflow through partially open inlets as is normal practice for the analysis of flow through sudden enlargements. If this is done the following results are obtained for subcritical inflow:

$$\frac{\dfrac{u'}{a'_{oINLET}}}{1 + \left(\dfrac{\gamma+1}{2}\right)\left(\dfrac{u'}{a'_{oINLET}}\right)^2} = \frac{\Phi\left(\dfrac{u'_{INLET}}{a'_{INLET}}\right)\left[1 + \dfrac{\gamma-1}{2}\left(\dfrac{u'_{INLET}}{a'_{INLET}}\right)^2\right]^{1/2}}{1 + \Phi\gamma\left(\dfrac{u'_{INLET}}{a'_{INLET}}\right)} \qquad (4.41)$$

and

$$\frac{\Phi\gamma\left(\dfrac{u'_{INLET}}{a'_{INLET}}\right)^2 + 1}{\left\{1 + \dfrac{\gamma-1}{2}\left(\dfrac{u'_{INLET}}{a'_{INLET}}\right)^2\right\}^{\gamma/(\gamma-1)}} = \frac{P'}{P'_{oINLET}}\left[\frac{1 + \dfrac{\gamma+1}{2}\left(\dfrac{u'}{a'_{oINLET}}\right)^2}{1 - \dfrac{\gamma-1}{2}\left(\dfrac{u'}{a'_{oINLET}}\right)^2}\right] \qquad (4.42)$$

Equations (4.41) and (4.42) take the place of Eq. (4.33) when it is desired to account for pressure recovery. The entropy relationship corresponding to Eqs. (4.41) and (4.42) is identical to that used when pressure recovery is neglected, namely Eq. (4.31). A possible iterative solution procedure for Eqs. (4.41), (4.42), and (4.31) requires a guess to be made of the unknown characteristic quantity $m'$ or $n'$ so that $u'$ can be evaluated. This allows $u'_{INLET}/a'_{INLET}$ to be established from Eq. (4.41). The value of $u'_{INLET}/a'_{INLET}$ is then substituted in Eq. (4.42) which permits $P'$ to be established. This value of $P'$ is, in turn, substituted in Eq. (4.31), together with the previously obtained value of $u'$, to evaluate $s'$. This then allows $P'$ to be evaluated from

$$P' = \left(\frac{m' + n'}{2}\right)^{2\gamma/(\gamma-1)} e^{-s'}$$

If this value of $P'$ differs from that obtained from (4.42) a revised value of $m'$, or $n'$, is used to replace the original guessed value and so on until convergence occurs.

The partly open inlet boundary condition with pressure recovery appears to have been originated by Cotter (1963). The only other difference between boundary conditions with pressure recovery and those applicable without pressure recovery relates to the choking condition expressed in terms of conditions at the boundary of the nonsteady flow field. In accordance with Cotter's results (Cotter, 1963) flow is choked at the inlet throat when:

$$P'_{oINLET} \geq \frac{(a')^{2\gamma/(\gamma-1)} e^{-s'}\left[1 + \dfrac{\gamma+1}{2}\left(\dfrac{u'}{a'_{oINLET}}\right)^2\right]\left(\dfrac{\gamma+1}{2}\right)^{\gamma/(\gamma-1)}}{(1 + \Phi\gamma)\left[1 - \dfrac{\gamma-1}{2}\left(\dfrac{u'}{a'_{oINLET}}\right)^2\right]} \qquad (4.43)$$

Equation (4.43) reduces to the result given by Eq. (4.26) when $\Phi = 1$. The analysis of inflow with supercritical pressure ratios is unaffected by considerations of pressure recovery since such a flow involves a free expansion where the flow leaves the inlet throat (see Fig. 4.5). This free expansion is incompatible with the classical analysis of a sudden enlargement. Hence the analysis of inflow with supercritical pressure ratios follows that presented in Section 4.4.3 with the exception that the appropriate choking condition is that given by Eq. (4.43).

## 4.5 Duct Branches

Branches are common in many duct systems where nonsteady flows can occur. Examples are the inlet and exhaust systems of multicylinder reciprocating engines, also natural gas distribution networks. If the branch takes the form of individual pipes communicating with a large, common vessel or plenum, as indicated in Fig. 4.6(a), then it is normally adequate to consider the possibility of nonsteady flow in each of the pipes connected to the plenum the average flow velocity within which is relatively low and hence can be ignored. In such cases the outflow and inflow conditions presented in Sections 4.3 and 4.4, respectively, are applicable. The additional, linking, boundary condition is that of continuity, i.e., at all times the rate of mass accumulation within the plenum must equal the net mass inflow rate into the plenum. The pressure and temperature within such a plenum are normally taken to be spatially uniform assumptions consistent with those of negligibly low flow velocities and a well-mixed flow.

For pipe branches where there is no significant volume enclosed at the junction of the pipes, as shown in Fig. 4.6(b), it is usual to assume, for nonsteady compressible flow, that the static pressure is equal, at the branch, in each of the pipes communicating with the branch (Benson et al., 1963; Benson, 1975). It is also interesting to note that an equivalent assumption was made by Streeter and Wylie (1981) for pipe branches in nonsteady incompressible flow.

The difficulties of implementing the constant static pressure boundary condition depend upon the sophistication required. Benson (1975) has produced a very simple algorithm for *homentropic* flow in a branch of any number, $n$, of joining pipes. Benson showed that when the requirements of continuity are taken into account and when:

$$K_r \equiv \frac{2A_r}{A_{\text{TOTAL}}} \tag{4.44}$$

and also:

$$A_{\text{TOTAL}} \equiv \sum_{r=1}^{n} A_r \tag{4.45}$$

it can then be shown that

$$\lambda_{\text{UNKNOWN}\,r} = \sum_{r=1}^{n} (K_r \lambda_{\text{KNOWN}\,r}) - \lambda_{\text{KNOWN}\,r} \tag{4.46}$$

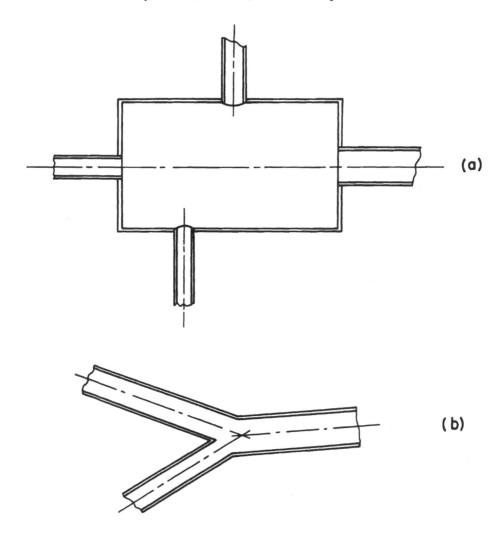

**Fig. 4–6** Pipe branches. (a) with plenum-like vessel; (b) without plenum.

where:

$A_r$ = cross-sectional area of the $r$th pipe communicating with the junction,

$\lambda_{\text{UNKNOWN}_r}$ = unknown, time dependent, characteristic quantity (i.e., $m'$ or $n'$) at the boundary of the $r$th pipe,

$\lambda_{\text{KNOWN}_r}$ = known, time dependent, characteristic quantity (i.e., $m'$ or $n'$) at the boundary of the $r$th pipe.

The $s'$ characteristic is ignored because only homentropic flow is considered (see Section 3.3.9). It is important to realize that in the notation of Benson's algorithm flow toward the junction is defined as having a positive velocity. It should also be noted that the possibility of heat exchange in the vicinity of the junction is not excluded. Heat exchange will be necessary, in principle at least, to counter the influences of any irreversibilities in the junction and hence maintain the uniform entropy required to justify a homentropic analysis of the junction flows.

### 4.6  Application of Boundary Condition Equations

It can be seen from this chapter that in a general sense application of all but the most simple boundary conditions, those for fully open or fully closed ends in particular, tends to be both tedious and relatively complicated. For the majority of nonsteady flow problems the selection and use of appropriate boundary condition equations add very substantially to the complexity of problem solving using the method-of-characteristics. To establish solutions rapidly it is virtually essential to use computerized numerical procedures.

However, manual problem solving is practical for some relatively simple problems. It is also feasible, with noncomputerized procedures, to make use of boundary condition charts that allow flows through partly open inlets, or outlets, to be handled with comparative ease. Both manual and computerized solution procedures are described in the next chapter.

### Problems

4.1  Show that the condition for subcritical outflow from an exit port is given by:

$$P'_{\text{EXIT}} \ge (a')^{2\gamma/(\gamma-1)} e^{-s'} \left\{ \frac{2}{\gamma+1} + \frac{\gamma-1}{\gamma+1} \left(\frac{u'}{a'}\right)^2 \right\}^{\gamma/(\gamma-1)}$$

4.2.  Show that for subcritical flow through a partly open exit in which the static pressure $\equiv P'_{\text{EXIT}}$:

$$\left(\frac{u'}{a'}\right)^2 = \frac{\dfrac{2}{\gamma-1}\left\{ 1 - (a')^{-2} \left(\dfrac{e^{-s'}}{P'_{\text{EXIT}}}\right)^{-(\gamma-1)/\gamma} \right\}}{\dfrac{a'^{4/(\gamma-1)}}{\Phi^2} \left(\dfrac{e^{-s'}}{P'_{\text{EXIT}}}\right)^{2/\gamma} - 1}$$

4.3  Show that, when the stagnation pressure in an outlet port is prescribed:

$$P'_{\text{oEXIT}} = (a')^{2\gamma/(\gamma-1)} e^{-s'} \left\{ 1 + \frac{\gamma-1}{2} \left(\frac{u'}{a'}\right)^2 \right\}^{\gamma/(\gamma-1)}$$

4.4  Show that, for partly open outlets, the boundary equation when the flow is supercritical can be written:

$$\frac{|u'|}{a'} = \Phi\left\{\frac{2}{\gamma + 1} + \frac{\gamma - 1}{\gamma + 1}\left(\frac{u'}{a'}\right)^2\right\}^{(\gamma+1)/2(\gamma-1)}$$

4.5 Show that the condition for subcritical flow through an inlet is given by:

$$P'_{\text{oINLET}} < \left(\frac{\gamma + 1}{2}\right)^{\gamma/(\gamma-1)}(a')^{2\gamma/(\gamma-1)}e^{-s'}$$

4.6 Show that for an inlet operating at subcritical conditions:

$$s' = \ln\left[\frac{P'_{\text{oINLET}}}{P'}\left\{1 - \frac{\gamma - 1}{2}\left(\frac{u'}{a'_{\text{oINLET}}}\right)^2\right\}^{\gamma/(\gamma-1)}\right] + s'_{\text{oINLET}}$$

4.7 Demonstrate that for a partly open inlet subjected to subcritical flow conditions:

$$\frac{\dfrac{|u'|}{a'_{\text{oINLET}}}}{1 - \dfrac{\gamma - 1}{2}\left(\dfrac{u'}{a'_{\text{oINLET}}}\right)^2} = \Phi\left(\frac{P'_{\text{oINLET}}}{P'}\right)^{(\gamma-1)/2\gamma}\sqrt{\frac{2}{\gamma - 1}\left\{\left(\frac{P'_{\text{oINLET}}}{P'}\right)^{(\gamma-1)/\gamma} - 1\right\}}$$

4.8 Show that the boundary conditions for a partly open inlet in which the flow is choked are given by:

i. $s' = \ln\left[\dfrac{1}{\Phi}\left(\dfrac{\gamma + 1}{2}\right)^{(\gamma+1)/2(\gamma-1)}\dfrac{|u'|}{a'_{\text{oINLET}}}\left\{1 - \dfrac{\gamma - 1}{2}\left(\dfrac{u'}{a'_{\text{oINLET}}}\right)^2\right\}^{1/(\gamma-1)}\right]$

$\quad + s'_{\text{oINLET}}$

and

ii. $\dfrac{\dfrac{|u'|}{a'_{\text{oINLET}}}}{1 - \dfrac{\gamma - 1}{2}\left(\dfrac{u'}{a'_{\text{oINLET}}}\right)^2} = \Phi\left(\dfrac{2}{\gamma + 1}\right)^{(\gamma+1)/2(\gamma-1)}\dfrac{P'_{\text{oINLET}}}{(a')^{2\gamma/(\gamma-1)}e^{-s'}}$

# 5

## SOLUTION TECHNIQUES

Two techniques are commonly used. A computerized procedure for solving the full equations of Chapters 3 and 4 is the only feasible one unless numerous simplifying assumptions are made. Hand calculations are quite practical, although slow, when it is adequate to consider isentropic flow in the flow field as a whole or in a small number of discrete, abutting, regions within each of which the flow is isentropic.

### 5.1 Graphical Solution Procedure

When a computer is not available it is quite practical to carry out calculations by semigraphical means, provided the flow field occurs in a duct of uniform cross-sectional area and is isentropic or can be divided into a small number of individual isentropic regions. For problems of a more elaborate nature in which friction and heat transfer, also entropy gradients, are present, particularly when the duct cross-sectional area is nonuniform, the semigraphical method is inordinately slow.

In the writer's opinion it is worthwhile spending some time obtaining solutions by semigraphical means to a few suitable problems for the purpose of gaining a deeper physical understanding of unsteady flow.

### 5.1.1 General technique

In general, the semigraphical procedure requires the simultaneous construction of two diagrams. One, termed the *wave diagram*, is drawn on the $x' \sim t'$ plane; the other, which is a service diagram to assist in the construction of the wave diagram, is called a *state diagram* and is constructed on the $u' \sim a'$ plane. Two simple examples utilizing this concept were presented in Chapter 3, Section 3.4.

The wave diagram is, for all practical purposes, the "answer" to the problem in that it contains information suitable for answering specific questions about the flow field; it also illustrates the flow field as a whole in terms of the characteristic pattern and, provided they have been added to the wave diagram, particle path lines.

In view of the nature of the $u' \sim a'$ construction it is possible to produce a generalized, master, $u' \sim a'$ (or state) diagram applicable to many problems. As had been shown previously, it is possible on the $u' \sim a'$, or state, plane to represent all possible state changes associated with isentropic waves by means

81

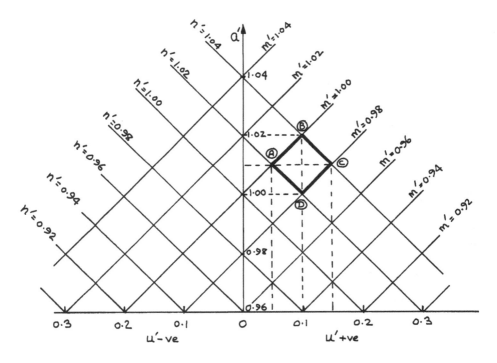

**Fig. 5–1** A state diagram drawn for equal increments of $m'$ and $n'$ of 0.02 ($\gamma = 1.4$).

of lines running at $\pm 45°$ (provided the ordinate scale is $2/(\gamma - 1)$ times that of the abscissa). By drawing families of such lines at small, but discrete, intervals a pattern is obtained as shown in Fig. 5.1; a similar diagram appeared in Chapter 2 (Fig. 2.6).

Other essential information that can be added to a generalized state diagram are the local slopes of the $m'$ and $n'$ lines on the $x' \sim t'$ plane. In practice it is easiest to employ the reciprocal slopes in preference to the slopes themselves. From Eqs. (3.60) and (3.61) the local reciprocal slope $\beta'_m$ of an $m'$ characteristic is $(u' - a')$ and that for an $n'$ characteristic is $(u' + a')$. Thus for the points marked $A$, $B$, $C$, and $D$ in Fig. 5.1:

Reciprocal slope of $m' = 1.00$ at $A = -0.96$ ⎫ Average between
Reciprocal slope of $m' = 1.00$ at $B = -0.92$ ⎭ $A$ and $B = -0.94$

Reciprocal slope of $m' = 0.98$ at $D = -0.90$ ⎫ Average between
Reciprocal slope of $m' = 0.98$ at $C = -0.86$ ⎭ $D$ and $C = -0.88$

Reciprocal slope of $n' = 1.02$ at $A = +1.06$ ⎫ Average between
Reciprocal slope of $n' = 1.02$ at $D = +1.10$ ⎭ $A$ and $D = +1.08$

Reciprocal slope of $n' = 1.04$ at $B = +1.12$ } Average between
Reciprocal slope of $n' = 1.04$ at $C = +1.16$ } $B$ and $C = +1.14$

By proceeding in this manner all the localized average reciprocal slopes can be written on the appropriate $m'$ and $n'$ lines on the state diagram.

Figure 5.2 (also see pocket in rear cover), a generalized form of state diagram for $\gamma = 1.4$, has been produced in the manner described; all of the localized average reciprocal slopes have been written on the $m'$ and $n'$ lines. Examination of the diagram shows that $u'$ positive is on the *left*. The purpose of this is simply to make the $n'$ characteristics slope upward to the right, and the $m'$ ones upward to the left in the state diagram. This means that the $m'$ and $n'$ lines on the state diagram ran in the same general directions as on the $x' \sim t'$, or wave, diagram when flow is subsonic.

A more detailed inspection of Fig. 5.2 shows that:

i. apart from reversal of signs between the left and right-hand portions, the state diagram is symmetrically disposed about the $a'$ axis,

ii. the numerical increments between adjacent localized average reciprocal slopes are equal—this would not have been the case had slopes rather than reciprocal slopes been written on the characteristics. It also justifies the process of performing an arithmetic averaging of the reciprocal slopes at $A$ and $B$, etc., are was done for the data applicable to Fig. 5.1.

The generalized $u' \sim a'$ diagram, Fig. 5.2, is adequate for the solution of many isentropic nonsteady flow problems in which the boundary conditions (at $x' = x'_{\text{MIN}}$ and $x' = x'_{\text{MAX}}$ for all $t'$) are either that the duct ends are:

i. closed
or
ii. fully open.

The boundary condition, $u' = 0$ for a closed end corresponds to, from Eq. (4.2):

$$m' = n' \qquad (5.1)$$

This is already in a form suitable for use in conjunction with a $u' \sim a'$ diagram.

The duct-end fully open boundary condition can, conveniently, be specified either on the basis of a prescribed static or a prescribed stagnation pressure. Since here we are considering isentropic flow it is not necessary to specify the entropy during filling as the entropy of the incoming fluid is, by implication, the same as that in the duct or cells.

Since the flow is isentropic, for a specified static pressure with subcritical flow (the condition of most general interest):

$$a'_{\text{BOUNDARY}} = (P'_{\text{BOUNDARY}})^{(\gamma - 1)/2\gamma} \qquad (5.2)$$

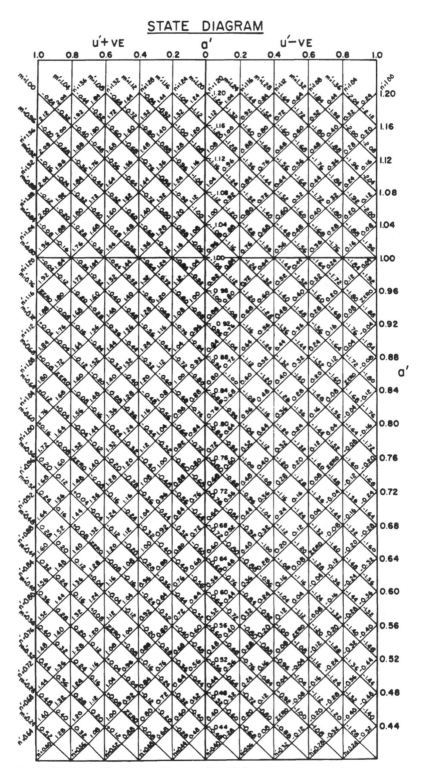

**Fig. 5-2** Generalized state diagram ($\gamma = 1.4$).

and when the stagnation pressure is specified, for subcritical flow

$$a'_{o\,\text{BOUNDARY}} = (P'_{o\,\text{BOUNDARY}})^{(\gamma-1)/2\gamma} \tag{5.3}$$

Thus on a boundary with a specified static pressure, the boundary condition in terms suitable for the $u' \sim a'$ diagram becomes:

$$a'_{\text{BOUNDARY}} = a' = \frac{m' + n'}{2} \tag{5.4}$$

and when the stagnation pressure is specified, invoking the energy equation allows the boundary condition to be expressed in terms suitable for representation on the $u' \sim a'$ plane:

$$(a'_{o\,\text{BOUNDARY}})^2 = a'^2 + \frac{\gamma-1}{2}u'^2 \tag{5.5}$$

Curves derived from Eqs. (5.4) and (5.5) can be added to the $u' \sim a'$ diagram as appropriate for cases involving subcritical isentropic inflow, or outflow, through fully open duct ends.

More complicated boundary conditions in which the cell ends are partly open require the introduction of additional charts based on a graphical presentation of some of the equations of Chapter 4. Before studying these additional diagrams and their use two examples will be presented for which only the $u' \sim a'$ diagram is required.

### 5.1.2   Illustrative examples

Examples of calculations carried out using only the $u' \sim a'$ diagram imply, for duct flows, that the duct ends are either fully open or fully closed. The physical interpretation of this is that any valves are located at the duct ends and such valves are, therefore, opened or closed instantaneously. Viscous effects are, of course, absent because flow is assumed to be isentropic.

### Example I

A duct containing fluid at rest, at a pressure exceeding that of the surroundings, is provided with a valve at the right-hand end; the left hand is closed at all times. The surroundings pressure is maintained at a constant (uniform) value. The initial duct pressure, $P_1$, divided by the surroundings pressure, $P_2$, $= 1.714$; $\gamma = 1.4$. Establish:

   i. the wave diagram indicating the maximum time for which the valve can remain open before inflow sets in,
   ii. the lowest pressure, as a fraction of the surroundings pressure, attained in the duct during the emptying process.

Choosing the static conditions in the fully open outlet as convenient reference conditions:

$$a'_1 = (1.714)^{1/7} = 1.08$$

and, from Eq. (5.2), the open end boundary condition is given by

$$a'_{\text{BOUNDARY}} = 1^{1/7} = 1.0$$

The closed-end boundary condition is given by Eq. (5.1), namely:

$$n' = m'$$

The boundary condition describing the duct initial conditions is also of the form

$$n' = m'$$

since the fluid in the duct is initially at rest.

Assuming that the valve opens instantaneously at $t' = 0$ and, since the valve is at the right-hand end of the duct, outflow occurs in the positive direction starting at $t' = 0$. Construction of the $x' \sim t'$ diagram (Fig. 5.3(a)) can now be

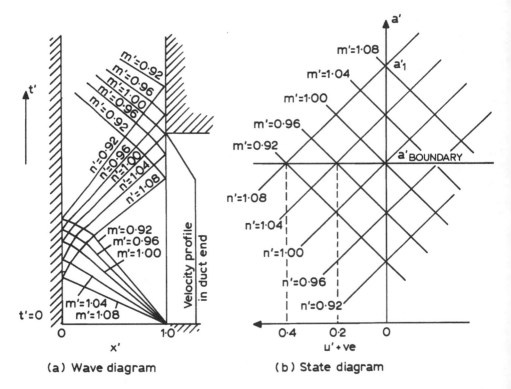

(a) Wave diagram            (b) State diagram

**Fig. 5–3** Wave and state diagrams for Example I.

commenced. (*Note*: Had the valve been located at the left-hand end of the duct, thereby giving a negative outflow direction, this would not alter the answer to the second part of the question.)

At the same time that the wave diagram is under construction, the emptying process must be traced out on the $u' \sim a'$ diagram (Fig. 5.3(b)).

The construction of the wave diagram is initiated at the corner $t' = 0$, $x' = 1.0$; a family of $m'$ characteristics propagates back into the duct. The first characteristic is $m' = 1.08$ its reciprocal slope is that of an $m' = 1.08$ characteristic crossing a region in which $n' = 1.08$. Thus from the $u' \sim a'$ diagram the reciprocal slope on the $x' \sim t'$ plane of the $m' = 1.08$ characteristic is

$$\frac{-1.04 - 1.12}{2} = -1.08$$

The second $m'$ characteristic is $m' = 1.04$; it also crosses the $n' = 1.08$ region (i.e., no $n'$ characteristics have yet been generated) and its reciprocal slope is

$$\frac{-0.92 - 1.00}{2} = -0.96$$

and so on for the $m' = 1.00$ and $m' = 0.96$ characteristics. It should be noted that the $m' = 0.92$ and $n' = 1.08$ lines of the state diagram intersect at $a' = a'_{\text{BOUNDARY}} = 1.0$. This means that the $m' = 0.92$ characteristic is the last member of the initial fan of $m'$ characteristics; its reciprocal slope is

$$\frac{-0.56 - 0.64}{2} = -0.60$$

When the $m' = 1.08$ characteristic reflects at the closed end it produces an $n' = 1.08$ characteristic ($n' = m'$ at the closed end). The $n' = 1.08$ characteristic runs between the $m' = 1.08$ and $m' = 1.04$ characteristics. Its reciprocal slope is, therefore (directly from the state diagram) $+ 1.12$.

The $m' = 1.04$ characteristic then travels between the $n' = 1.08$ characteristic and the closed end from which it reflects as an $n' = 1.04$. The reciprocal slope of the $m' = 1.04$ between the $n' = 1.08$ and $n' = 1.04$ characteristics is, therefore, $-1.00$.

The construction of the meshed portion of the wave diagram, near the closed end, proceeds in the foregoing manner until reflection of the $m' = 0.92$ characteristic has occurred at the closed end. The $n' = 0.92$ to $n' = 1.08$ characteristics have reciprocal slopes commensurate with the crossing of an $m' = 0.92$ space in the region between the last member ($m' = 0.92$) of the initial expansion fan and the first $m'$ characteristic reflected from the open port. Attention can now be focused on the open end of the duct.

The $n' = 1.08$ characteristic reflects at the end (boundary condition $a'_{\text{BOUNDARY}} = 1.00$) as an $m' = 0.92$: this is consistent with Eq. (5.4). The $m' = 0.92$ characteristic has a slope of $-0.64$ between the $n' = 1.08$ and

$n' = 1.04$ characteristics. The outflow velocity $u'$ remains at the initial value of 0.4 up to and including the point at which the $n' = 1.08$ reflects as an $m' = 0.92$. The $n' = 1.04$ reflects as an $m' = 0.96$, however, the intersection of the $m' = 0.96$ and $n' = 1.04$ characteristics with the $a'_{\text{BOUNDARY}} = 1.00$ line occurs at $U' = 0.2$. The $n' = 1.00$ characteristic reflects as an $m' = 1.00$; these characteristics intersect the $a'_{\text{BOUNDARY}} = 1.00$ line at $u' = 0$. The valve must, therefore, be closed at this instant if inflow is to be prevented.

The $n' = 0.96$ and $n' = 0.92$ characteristics reflect from the closed end as $m' = 0.96$ and $m' = 0.92$ characteristics, respectively.

*Answers*

   i. the time for which the valve may remain open before inflow occurs can be read from the wave diagram. This could be converted from normalized to conventional time had sufficient data for this purpose been given in the question.
   ii. the lowest value of $a'$ occurs at the left-hand closed end in the region enclosed between the $n' = m' = 0.92$ characteristics. Thus:

$$P'_{\text{DUCT MIN}} = (0.92)^7 = 0.558$$

*Example II*

Draw the wave diagram for a shock tube, of uniform cross-sectional area, with both ends sealed off. A diaphragm is located halfway along the length of the tube. The fluid at both sides of the diaphragm is air ($\gamma = 1.4$); the fluid is initially at rest and the pressure ratio across the diaphragm is 3.58. The entropies in the high and low pressure regions are equal.

The high pressure region is to the left of the diaphragm; take the reference conditions to be those prevailing initially in the low-pressure region. The diaphragm is ruptured at $t' = 0$. Add to the wave diagram the particle path line separating the contents of the initial high and low-pressure zones.

Since the initial pressure ratio, $P_{\text{HP}}/P_{\text{LP}} = 3.58$ and also because the entropies are uniform throughout:

$$a'_{\text{HP INITIAL}} = (3.58)^{1/7} = 1.2$$

and

$$a'_{\text{LP INITIAL}} = 1.00$$

Construction of the wave diagram can now proceed. The wave diagram and the portion of the state diagram required for establishing the state changes associated with compression waves propagated in the region to the right of the diaphragm, after diaphragm rupture, are shown in Fig. 5.4.

The construction of the wave diagram in the region $0 \leq x' \leq 0.5$ requires no explanation being similar to that of the previous example.

An attempt to apply the principles explained in detail in the previous example to the flow in the region $0.5 \leq x' \leq 1.0$ leads to difficulties in that the

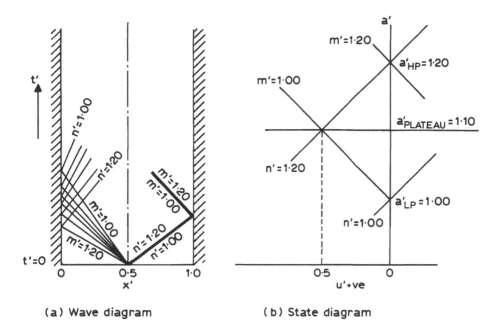

(a) Wave diagram                 (b) State diagram

**Fig. 5–4** Wave and state diagrams for Example II.

sequence of the $n'$ characteristics is such that they attempt to overtake each other. This situation implies that a shock wave has been formed. Strictly this case should be handled by application of the normal shock relations to the moving wave. In order to avoid this complication it is usually acceptable, for fairly weak shocks, to treat the shock wave as a bundle of characteristics, in this case, because the shock is propagating to the right, as $n'$ characteristics. The reciprocal slope assigned to the shock wave comprising $n'$ characteristics from $n' = 1.00$ to $n' = 1.20$ inclusive is the average of the individual reciprocal slopes of the characteristics making up the shock wave. In this case the value obtained from the state diagram (the shock crosses an $m' = 1.00$ space) is $+1.30$. The $n'$ shock is found ($n' = m'$ at a closed end) to reflect from the closed end as an $m'$ shock; the reciprocal slope of the reflected shock is, from the state diagram, $-0.90$.

A clue to the construction of the wave diagram is that the velocity and the static pressure of the fluid in the "plateau" region between the foot of the expansion wave and the crest of the shock wave ($m' = 1.0$ and $n' = 1.20$ characteristics) is the same whether arrived at by the expansion of the high pressure, or compression of the low pressure, fluid. This would, of course, also have been the case had the entropies on each side of the diaphragm not been equal resulting in an entropy discontinuity.

### 5.1.3  Entropy discontinuity

A characteristic reaching an interface between two regions having different entropies is, in general, both reflected and transmitted at the entropy interface. An example of such a situation is shown in Fig. 5.5.

With reference to the notation of Fig. 5.5, from the uniformity of static pressure across the entropy interface:

$$P'_A = P'_B \tag{5.6}$$

$$P'_C = P'_D \tag{5.7}$$

and from the uniformity of fluid velocity at the interface:

$$u'_A = u'_B \tag{5.8}$$

$$u'_C = u'_D \tag{5.9}$$

Expressing (5.7) and (5.9) in terms of $n' \cdot m'$, and $s'$ from Eqs. (3.68), (3.69), and (3.72) (Chapter 3), the following simultaneous equations are obtained:

$$\left(\frac{n'_C + m'_C}{2}\right)^{(2\gamma/(\gamma-1))_{\mathrm{I}}} e^{-S_{\mathrm{I}}} = \left(\frac{n'_D + m'_D}{2}\right)^{(2\gamma/(\gamma-1))_{\mathrm{II}}} e^{-S_{\mathrm{II}}} \tag{5.10}$$

$$\frac{n'_C - m'_C}{(\gamma-1)_{\mathrm{I}}} = \frac{n'_D - m'_D}{(\gamma-1)_{\mathrm{II}}} \tag{5.11}$$

If the incident wave is of the $n'$ family and (therefore) comes from the left, as shown in Fig. 5.5, then

$$n'_C = n'_X \tag{5.12}$$

$$m'_D = m'_B \tag{5.13}$$

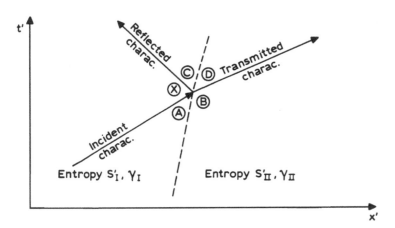

**Fig. 5–5** Incident, transmitted, and reflected waves at an entropy discontinuity.

Substitution in (5.10) and (5.11) for $n'_C$ and $m'_D$ in terms of $n'_X$ and $m'_B$, which are both known, yields

$$\left(\frac{n'_X + m'_C}{2}\right)^{(2\gamma/(\gamma-1))_{\mathrm{I}}} e^{-S_{\mathrm{I}}} = \left(\frac{n'_D + m'_B}{2}\right)^{(2\gamma/(\gamma-1))_{\mathrm{II}}} e^{-S_{\mathrm{II}}} \qquad (5.14)$$

$$\frac{n'_X - m'_C}{(\gamma - 1)_{\mathrm{I}}} = \frac{n'_D - m'_B}{(\gamma - 1)_{\mathrm{II}}} \qquad (5.15)$$

Equations (5.14) and (5.15) can be solved simultaneously for the unknowns $m'_C$ and $n'_D$. For particular cases where $\gamma_{\mathrm{I}} = \gamma_{\mathrm{II}}$ an entropy discontinuity will only arise due to a temperature discontinuity. This point can be demonstrated from application of Eq. (3.72) in regions I and II prior to the arrival of the incident wave:

$$P'_{\mathrm{I}} = (a'_{\mathrm{I}})^{(2\gamma/(\gamma-1))_{\mathrm{I}}} e^{-S'_{\mathrm{I}}}$$

and

$$P'_{\mathrm{II}} = (a'_{\mathrm{II}})^{(2\gamma/(\gamma-1))} e^{-S'_{\mathrm{II}}}$$

hence since $P'_{\mathrm{I}} = P'_{\mathrm{II}}$ (using a common reference pressure, $P_{\mathrm{REF}}$, in regions I and II) and $\gamma_{\mathrm{I}} = \gamma_{\mathrm{II}} = \gamma$:

$$\left(\frac{a'_{\mathrm{I}}}{a'_{\mathrm{II}}}\right)^{2\gamma/(\gamma-1)} = e^{(S'_{\mathrm{I}} - S'_{\mathrm{II}})}$$

or when $T_{\mathrm{REF}}$ is also common to both regions:

$$\frac{T_{\mathrm{I}}}{T_{\mathrm{II}}} = \{e^{(S_{\mathrm{I}} - S_{\mathrm{II}})}\}^{\gamma/(\gamma-1)} = \frac{T_A}{T_B} \qquad (5.16)$$

It is, in fact, vital that common reference conditions, $P_{\mathrm{REF}}$ and $T_{\mathrm{REF}}$, be used in the regions on either side of an entropy discontinuity for the foregoing treatment of an entropy discontinuity to be valid.

Clearly a similar analysis can be performed for an $m'$ incident characteristic arriving from the right. As can be perceived from the considerations presented here the introduction of even a single entropy discontinuity considerably slows down the noncomputerized prediction procedure.

Under the special circumstances, known as the impedance-matched conditions, it is possible for an entropy discontinuity to be transparent to an incident wave. In such cases there is no reflection at the interface.

### 5.1.4  Impedance matching

Figure 5.6 shows an impedance-matched situation in which an incident wave, in the example illustrated an $n'$ wave, is transmitted through an entropy discontinuity without reflection occurring. This case can be analysed using Eq. (5.9), namely:

$$u'_C = u'_D$$

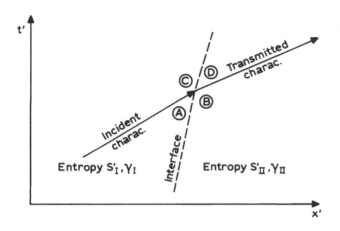

**Fig. 5–6** Transmittal, without reflection, of an incident wave crossing an entropy discontinuity (Impedance Matching).

and the implication that a reflected $m'$ wave is absent:

$$m'_C = m'_A \qquad (5.17)$$

in conjunction with Eq. (5.13):

$$m'_D = m'_B$$

Invoking Eq. (3.68) and substituting from (5.9), (5.13), and (5.17):

$$\frac{n'_C - m'_A}{(\gamma - 1)_{\mathrm{I}}} = \frac{n'_D - m'_B}{(\gamma - 1)_{\mathrm{II}}}$$

hence

$$n'_D = \frac{(\gamma - 1)_{\mathrm{II}}}{(\gamma - 1)_{\mathrm{I}}} \{n'_C - m'_A\} + m'_B \qquad (5.18)$$

Here $n'_D$ can be evaluated explicitly from (5.18) provided that full knowledge of the flow conditions prevails in both regions prior to the arrival of the incident wave of known strength.

It is important to realize that an essential condition applicable to both regions is that before arrival of a weak (acoustic) incident wave the acoustic impedances are equal in both flows; this condition is given by

$$\rho_{\mathrm{I}} a_{\mathrm{I}} = \rho_{\mathrm{II}} a_{\mathrm{II}} \qquad (5.19)$$

It is easy to show that (5.19) reduces to the equality:

$$\frac{\gamma_{\mathrm{I}}}{a_{\mathrm{I}}} = \frac{\gamma_{\mathrm{II}}}{a_{\mathrm{II}}} \qquad (5.20)$$

It can be seen, from (5.20), that when $\gamma_I = \gamma_{II}$ impedance matching corresponds to the trivial condition:

$$a_I = a_{II}$$

or

$$T_I = T_{II}$$

and hence, as is confirmed by Eq. (5.16), an entropy discontinuity does not exist.

For cases involving strong incident waves the impedance-matching condition is more complex than that corresponding to Eqs. (5.19) or (5.20). However, such cases are most likely to be dealt with by the use of numerical methods and hence they will not be considered further here. A more detailed treatment of impedance matching is available (Rudinger, 1969).

An incident $m'$ wave impedance-matching problem can be handled in the same manner as for an incident $n'$ wave case.

### 5.1.5   Partially open ends

These can be handled by presenting the appropriate Jenny boundary conditions, previously referred to in Chapter 4 (Jenny, 1950), or indeed any other boundary conditions, for example Cotter's filling process boundary conditions (Cotter, 1963), for partly open ends in terms of $\Phi$, $u'$, and $a'$. Curves of constant $\Phi$ can then be superimposed upon the $u' \sim a'$ diagram. This idea is illustrated in sketch form, in Fig. 5.7.

In practice, it is not convenient to add the filling and emptying boundary curves to the $u' \sim a'$ diagram and, hence, they are drawn separately because:

i. the reference conditions for the $u' \sim a'$ diagrams cannot, in general, be those for all the flow conditions for all possible situations. This means that boundary curves would have to be added at the appropriate locations on the $u' \sim a'$ diagram to suit each particular problem.

ii. the scale of the boundary curves is a function of the reference condition employed—this prevents the overlaying of a set of movable, transparent, master boundary curves as a solution to problem (i).

The use of boundary curves is best illustrated by means of an example. One example will suffice since the principles involved are the same for the opening, or closing, of an inlet or outlet valve at the end of a duct.

Consider Example I of Section 5.1.2. Reworking that example to take into account the finite time taken to open the outlet valve at the duct end is a straightforward application of the procedure. It is, of course, necessary to know the rate at which the valve is opened, and later closed, as a function of $t'$. Assuming a linear rate of opening a wave diagram of the form shown in Fig. 5.8 results.

By interpolating between the $\Phi$ lines on the state diagram with boundary curves superimposed, it can be seen for illustrative purposes, that the valve is

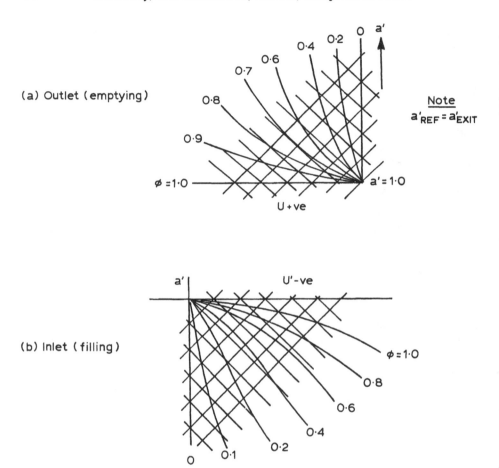

**Fig. 5–7** Sketch showing outlet and inlet boundary curves superimposed on the state plane.

approximately 17 percent of fully open when the $m' = 1.04$ characteristic starts. It is 36 and 62 percent open when the $m' = 1.00$ and $m' = 0.96$ characteristics start, respectively. The $m' = 0.92$ characteristic starts when the valve is fully open ($\Phi = 1.00$). The effect of the finite rate of opening of the valve is to spread out the wave fan initiating the emptying process.

A similar use of the boundary curves for filling processes implies neglect of the (small) entropy increase associated with filling processes when an inlet valve at a duct end is only partly open.

Figures 5.9 and 5.10 (also see pocket in rear cover) are the normalized Jenny boundary curves for filling and emptying. The normalization is with reference

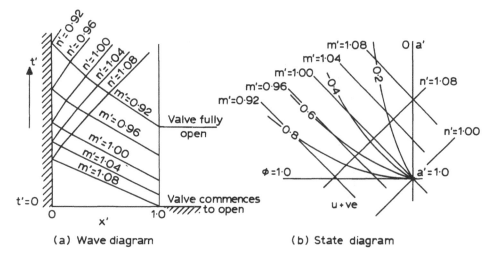

(a) Wave diagram           (b) State diagram

**Fig. 5–8** Sketch of wave and state diagrams for an emptying process with a finite rate of valve opening.

to the port stagnation and static conditions for inlets and outlets, respectively. Account must be taken of this when the $u' \sim a'$ diagram has been normalized by using other reference conditions.

The dimensionless mass flow contours on both the filling and emptying boundary curves can be of use when integrating the flow into, or out of, a duct.

For a duct the instantaneous dimensionless mass flow into, or out of, the valve can be integrated over the period for which the valve is partially, or fully, open in order to establish the mass of fluid added to, or removed from, the duct.

## 5.2 Numerical Solution Procedure

This employs the Hartree (1958) backward difference method and, consistently with the equations of Chapters 3 and 4, uses Eulerian ($x' \sim t'$) coordinates as distinct from Lagrangian coordinates in which the abscissa is directly proportional to fluid mass, and the ordinate to $t'$ as in the Eulerian system. The advantage of the Lagrangian system relates to simplification of the treatment of internal grid points. Particle paths (i.e., $s'$ characteristics) lie on ordinates of the diagram in the Lagrangian presentation. However, treatment of the boundary conditions appears to be more complicated in the Lagrangian system than in the Eulerian thereby canceling out the advantage previously referred to. Furthermore, it is less easy to allocate computer storage space efficiently when a mass $\sim t'$ (Lagrangian) system of coordinates is used.

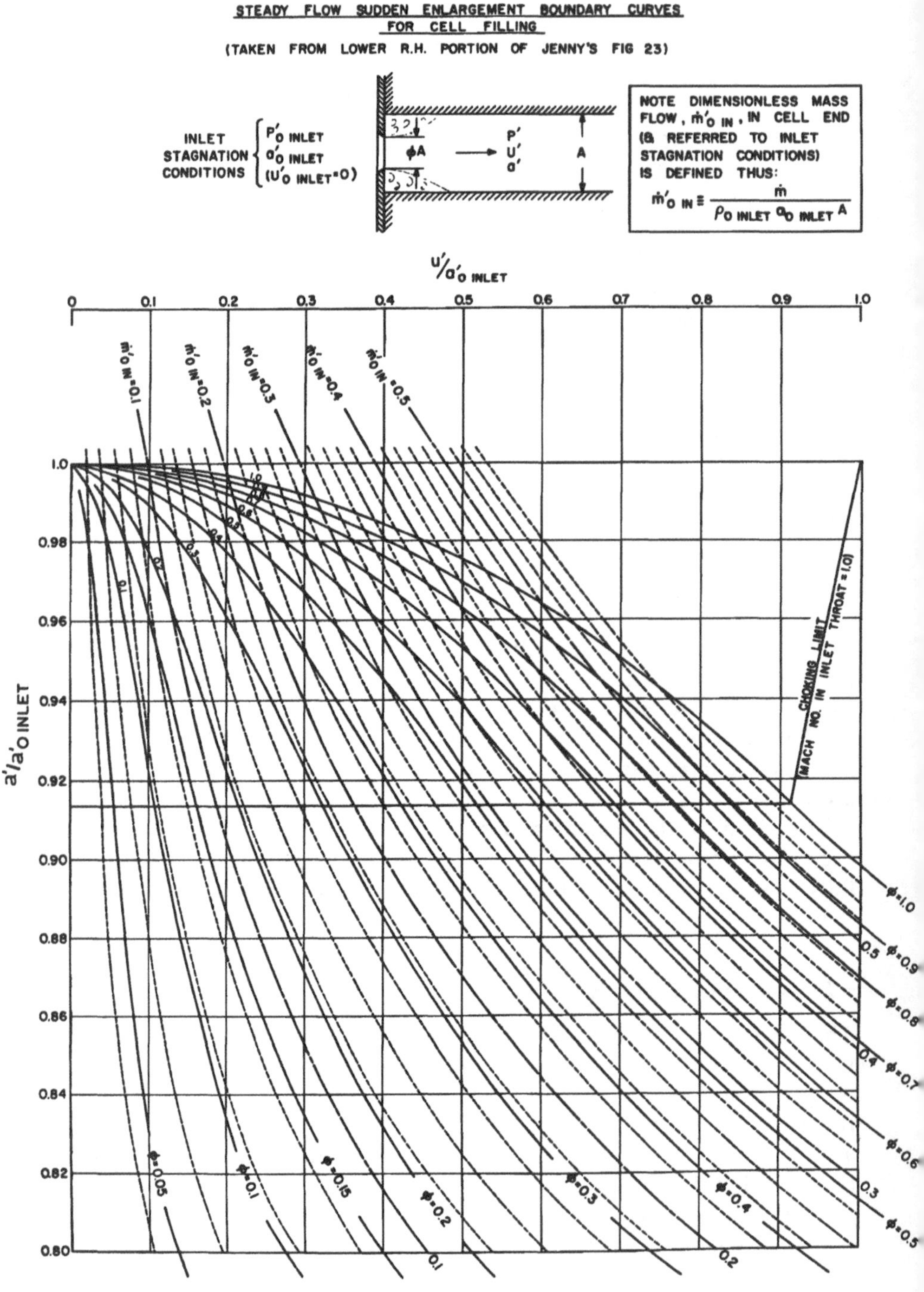

Fig. 5–9 Normalized boundary curves for filling processes ($\gamma = 1.4$) (after Jenny).

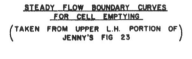

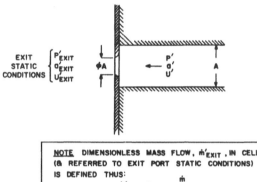

NOTE DIMENSIONLESS MASS FLOW, $\dot{m}'_{EXIT}$, IN CELL END
(& REFERRED TO EXIT PORT STATIC CONDITIONS)
IS DEFINED THUS:

$$\dot{m}'_{EXIT} = \frac{\dot{m}}{\rho_{EXIT}\, a_{EXIT}\, A}$$

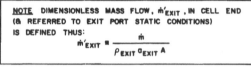

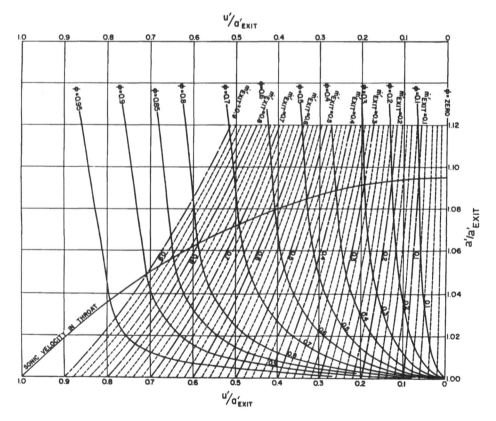

**Fig. 5–10** Normalized boundary curves for emptying processes ($\gamma = 1.4$) (after Jenny).

### 5.2.1   Internal nodes

In the Eulerian $(x' \sim t')$ coordinate system internal nodes, that is those not adjacent to boundaries, appear as shown in Fig. 5.11 that depicts a typical forward step employing a backward difference, and, for comparison, a typical forward step using a forward difference.

The purpose of the forward stepping procedure is to establish the values of the normalized characteristic quantities, $m'$, $n'$, and $s'$, at each of the nodal points at $t' + \Delta t'$ on the basis of their known values at $t'$.

The values of $n'$ at ① and $s'$ at ② are found by linearly interpolating between the known values at $x'_{(n-1)}$ and $x'_n$. Similarly, the value of $m'$ at ③ is also found by linear interpolation between known values of $m'$ at time $t'$; in this case the interpolation is between points $x'_n$ and $x'_{(n+1)}$.

Points ①, ②, and ③ are each located such that characteristics drawn through them pass through the point $x'_n$, $t' + \Delta t'$—labeled ④ in the sketch. Examination of Eqs. (3.63), (3.64), and (3.65) of Chapter 3 shows that in order to cope with the finite difference equivalent of the $\partial s'/\partial x'$ terms in (3.63) and (3.64), the entropy characteristic must be projected forward (i.e., from ② to ④ before either the $n'$ or $m'$ characteristics are advanced to point ④).

The slopes of each characteristic can, in the most simple treatment, be based

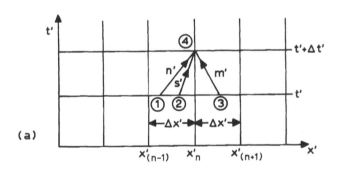

(a)

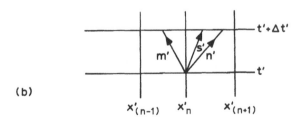

(b)

**Fig. 5–11** Forward stepping procedures for internal nodes, (a) Hartree backward difference scheme, (b) forward difference scheme.

on local conditions at points ①, ②, and ③ for the $n'$, $s'$, and $m'$ characteristics, respectively.

The maximum permissible value of $\Delta t'$ for any forward step is dictated by the requirement that points ①, ②, and ③, when related to point $x'_n$, must lie within the limits:

$$x'_{(n-1)} \leq x'_{①,② \text{ or } ③} \leq x'_{(n+1)}$$

i.e., the maximum $\Delta t'$ is that which makes the characteristic with the least slope a diagonal of the grid. A check should, therefore, be carried out within the program at each forward step to ensure that the above rule is not broken; $\Delta t'$ is thus a variable.

### 5.2.2   Boundary nodes

When a grid point lies on a boundary surface, the appropriate boundary condition is invoked to establish the required characteristic value. For example, when $x'_n$ is in the region of (say) an outlet port on the right information that would normally be derived from the $m'$ characteristic is absent as indicated in Fig. 5.12. Using the notation of Fig. 5.11 $m'$ at ④ is established from the appropriate outflow boundary condition of Chapter 4.

The situation for an inlet on the left is illustrated in Fig. 5.13. Here $n'$ and $s'$ are established from the appropriate inlet boundary conditions also presented in Chapter 4.

### 5.2.3   Smearing of wave pattern

The procedure described in Section 5.2 is not strictly in accordance with the ideals of the method-of-characteristics procedures although it is a fairly close model of them. Due to this, for example, the use of linear interpolation to establish the values of the characteristic quantities between nodal points, etc.,

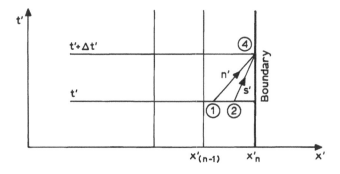

**Fig. 5–12** Grid scheme with an outlet located on the right-hand boundary.

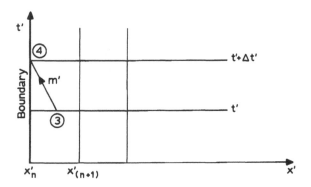

**Fig. 5–13** Grid scheme with an inlet located on the left-hand boundary.

the numerically obtained solutions tend to smear-out wave events. A compression, or expansion, wave will tend to be smeared-out, or smoothed, as indicated in Fig. 5.14.

The smearing effect has not, in practice, been found to be serious and does not eliminate anything but the most minor features of the flow picture.

### 5.2.4   Change of dependent variables

Instead of using $s'$ and the Riemann variables $m'$ and $n'$ alternative dependent variables are actually employed in the computer program. The reasons for the change of variables are entirely connected with improving the convergence of the computer program and are not fundamental to the analysis presented in Chapters 3 and 4, all of the main equations of which are employed, after being rewritten in terms of the new variables, in the computer program. The nature of the flow problem is such that $s'$, $m'$, and $n'$ remain, of course, the three characteristic quantities of interest; the change of variables merely means that $m'$, $n'$, and $s'$ are expressed as functions of the new dependent variables. The

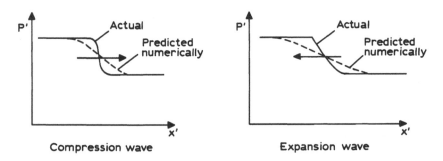

**Fig. 5–14** Sketch illustrating numerically induced smearing.

new dependent variables are:

$$U \equiv \left(\frac{\gamma - 1}{2}\right)u \tag{5.21}$$

$$\mathscr{P} \equiv P'^{(\gamma - 1)/2\gamma} \tag{5.22}$$

$$\chi \equiv e^{S'(\gamma - 1)/2\gamma} \tag{5.23}$$

Use of these variables, which were introduced by Spalding (1969), resulted in the eradication of numerical difficulties associated with the handling of the Riemann type characteristic quantities, $m'$ and $n'$, in the vicinity of steep entropy gradients.

A list of all the main equations of Chapters 3 and 4 in terms of the dependent variables $U$, $\mathscr{P}$, and $\chi$ has been given by Matthews (1969).

A fundamental version of the computer program applicable to ducts of uniform cross-sectional area has been described by Spalding (1969) and, inclusive of comparative results obtained using classical method-of-characteristics techniques, by Jonsson et al. (1973). A version of essentially the same program adapted to incorporate ducts with (spatially) variable cross-sectional area has been described in detail by Cronje (1979). An outline of the analytical procedure used by Cronje is also available (Cronje and Kentfield, 1980).

### 5.2.5 Finite difference formulation

In order to advance from knowledge of the new dependent variables $U$, $\mathscr{P}$, and $\chi$ at the nodal points of the grid at time $t'$ to those applicable at $t' + \Delta t'$ it is first necessary to establish, by linear interpolation, the values applicable at the feet of the characteristics at time $t'$. Referring to Fig. 5.11(a), which illustrates the backward difference forward stepping scheme incorporated in many computer programs, the slopes of the characteristics focusing on point ④ are based solely, at initiation of the forward stepping procedure, on the linearly interpolated values for points ①, ②, and ③.

It has been found, however, that this process is, for many problems, of insufficient accuracy and an improved accuracy is obtained by a re-evaluation utilizing data derived for nodal point ④ from the initial forward step previously described. Clearly the re-evaluation procedure can be repeated many times, if necessary, until data for nodal point ④ cease to be refined significantly by any subsequent re-evaluation.

In practice it is usually found that it is better, i.e., more accurate and more economical, to choose a relatively small spatial gap between adjacent nodes and to apply the re-evaluation procedure only once. Examples of equations used in a fully comprehensive backward difference forward stepping procedure have been listed by Cronje (1979).

### 5.2.6 Alternative numerical procedures

Methods differing from the one described, in detail, previously have been, and are, employed by some workers. An example is the so-called "nonmesh" technique of Benson et al. (1964) in which the characteristics do not focus at the nodal stations at $t' + \Delta t'$. The Benson et al. procedure is similar to, but not identical with, the forward difference scheme illustrated, diagrammatically, in Fig. 5.11(a).

Perhaps the most common of the alternative numerical procedures is the Lax-Wendroff method, a description of which is available (Lax and Wendroff, 1960). A paper on the application of boundary conditions to the Lax-Wendroff procedure, and a comparison of numerically predicted with corresponding exact solutions for several specific problems, is due to Warren (1983).

## Problems

NOTE. Assume, unless otherwise stated, that $\gamma = 1.4$.

5.1 Draw to scale the wave diagram for Example I (Section 5.1.2), and establish the period for which the valve remains open in terms of normalized time, $t'$, referred to static conditions in the open end of the duct and to the length of the duct. Also determine the initial displacement ($x' \equiv x/\text{LENGTH OF DUCT}$), from the left-hand end of the duct, of the last particle of fluid removed during outflow.

5.2 Construct the wave diagram for a shock tube of uniform cross-sectional area both ends of which are enclosed. The diaphragm is located half way along the tube; the entropies in the left-hand high-pressure and right-hand low-pressure sections are equal. The initial pressure ratio $P'_{\text{HP}}/P'_{\text{LP}} = 3.076$. The fluid in both sections is initially at rest. Use the conditions on the "plateau" between the expanded high pressure and compressed low pressure fluids for reference purposes. Plot the pressure distribution along the tube at time $t' = 1.2$. (Hint: The expansion fan in the left-hand section can be traced from question 5.1.)

5.3 Draw the wave diagram for an isentropic duct filling process in which the ratio of the initial duct pressure divided by the uniform inflow stagnation pressure = 0.35. Use the inflow stagnation conditions for reference purposes. The valve is opened, and closed, instantaneously.

Evaluate the ratio of the final duct pressure divided by the inflow stagnation pressure. Also determine the duration of the valve-open period, in terms of dimensionless time $t'$ and the inflow Mach number.

5.4 A weak shock-like compression wave coming from the right, as shown in Fig. 5.15, raises the value of $m'$ in region II by 0.1. The compression is reflected at, and transmitted through, the interface between region II and region I. The normalized entropy, $S_I$, in region I is zero. Evaluate $m'$ and $n'$ in the zones $A$, $B$, $C$, and $D$; also evaluate the resultant gas velocity $U'_C = U'_D$. Is the reflected wave a compression or an expansion? The adiabatic index $\gamma$ is equal in regions I and and II.

5.5 Draw the wave diagram for a duct emptying process in which the ratio of the initial duct pressure to the (uniform) pressure of the surroundings is

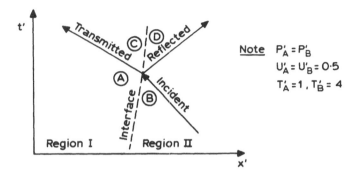

**Fig. 5–15** Diagram for problem 5.4.

1.714. The initial conditions in the duct are uniform and the fluid therein is at rest. The valve at one end of the duct requires a dimensionless time of $\Delta t' = 0.5$ to fully open and to fully close. The time interval, in $t'$ units, from the beginning of valve opening to the beginning of closure is 1.75. Use the static conditions of the outflow through the fully open valve for reference purposes.

   Plot the dimensionless mass flow, $\dot{m}'_{\text{EXIT}}$, passing through the valve as a function of $t'$. Is the port the widest possible if flow reversal is to be avoided? Is more, or less, fluid removed from the duct than for the emptying process of question 5.1?

5.6 A pipe, shown in Fig. 5.16, is filled with compressed air at rest and at a uniform temperature. A valve at one end of the pipe is opened gradually allowing the contents to communicate with the surroundings; the flow area through the valve increases linearly with time for a period of 5 seconds; during this period the effective flow area increases from zero to 60 percent of the pipe cross-sectional area. At 5 seconds after commencing to open the valve is shut instantaneously. Establish the pressure distribution in the pipe 10 seconds after commencing to open the valve. The ratio of the initial pressure in the pipe to the surroundings pressure is 1.714. Use the acoustic

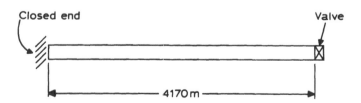

**Fig. 5–16** Diagram for problem 5.6.

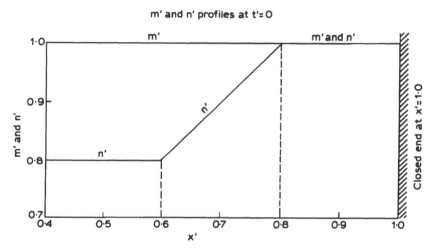

**Fig. 5–17** Diagram for problem 5.7.

velocity in the pipe at the initial conditions, 335 m/s, and the length of the pipe for reference purposes. Assume isentropic flow within the pipe.

What is the ratio of the final pressure in the pipe to the surroundings pressure when all the residual motion has died away? Assume that the final and initial temperatures of the contents of the pipe are equal.

5.7 Act as a human digital computer and perform, for the initial (i.e., $t' = 0$) $m'$ and $n'$ profiles shown in Fig. 5.17, four numerical forward steps by the backward difference (Hartree, 1958) technique described in Section 5.2. Compare the final $m'$ and $n'$ profiles with those obtained from conventional method-of-characteristics hand computations for isentropic flow. Make each numerical step for $\Delta t' = 0.05$; let $\Delta x' = 0.10$, where:

$$t' \equiv \frac{a_{(m'=n'=1.0)}t}{l}, \qquad x' \equiv \frac{x}{l}$$

Test that $\Delta t' = 0.05$ is within the maximum allowable limit at each forward step for the prescribed value of $\Delta x'$. Perform the Hartree calculations by graphical means.

# 6

## INITIAL-VALUE PROBLEMS

Shock tubes and gas pipelines are typical examples of noncyclic applications of (in essence) one-space-dimensional, time-dependent, compressible-flow theory. Such problems demand that the initial conditions, applicable at the beginning of the solution process, be specified correctly. This is in contrast to the corresponding situation for cyclic phenomena where it is essential, for demonstrating a converged solution, that the final conditions at the end and the beginning of a cycle are identical. Since these conditions are not normally known in advance it is quite usual, for cyclic processes, to assume simple nonexact, but preferably realistic, initial conditions along the "time zero" boundary.

It is, of course, essential that for both initial value and cyclic type problems the spatially defined boundary conditions be expressed correctly. Normally, but not exclusively, such conditions apply at the ends of the system (i.e., at $x' = 0$, and $x' = 1.0$). For gas pipelines in particular it is quite commonplace for valves, etc., to be located at stations other than, or in addition to, $x' = 0$ and $x' = 1.0$. For shock tubes it is usual for the frangible diaphragm separating the high and low-pressure zones to be located between the $x' = 0$ and $x' = 1.0$ stations. However, it is normally the complete and, it is assumed, instantaneous rupture of the diaphragm at time zero that initiates the flow field within a shock tube.

### 6.1 Shock Tubes

Shock tubes usually serve as research devices to generate, by very simple means, both a moving shock wave that passes over a model, or a material, under test and a high velocity flow that prevails, for a very short period of time, in the flow field behind the shock wave. The most simple version of a shock tube consists of a pipe, or tube, of uniform cross-sectional area closed at each end. A thin diaphragm separates the tube, diametrically, into two sections. One portion of the tube is pressurized and, if very high flow Mach numbers or shock strengths are required, the other portion is evacuated. When the desired initial pressure ratio has been achieved the tube is sealed off from the pressurization and evacuation systems and the diaphragm is ruptured, usually mechanically by means of a spike, to initiate the flow. The surface of the diaphragm is often prescored to facilitate rupture.

For most test work very large initial pressure ratos prevail between the high

105

and low-pressure sections of shock tubes: In some cases each section of a shock tube is charged with a different gas. Shock tubes operated under more modest conditions, with relatively low initial pressure ratios, lend themselves as very suitable devices for comparing, often with experiment, the results of various analytical procedures for predicting one-space-dimensional nonsteady, compressible, flow fields. Largely because of the noncyclic nature of the flow field in a shock tube, and also because of the well-defined initial conditions, it is comparatively easy to establish, by hand calculation, the predicted performance employing the full method-of-characteristics in conjunction with treatment of the initial and reflected compression waves as moving normal shocks.

### 6.1.1   Comparison of solutions

The upper portion of Fig. 6.1 shows a simple, hypothetical, shock tube in which the diaphragm is located at one-third of the overall length of the device from the left-hand end. The lower portion of Fig. 6.1 is a sketch, on the $t'$ versus $x'$ plane, of the wave events occurring following instantaneous rupture of the diaphragm. Figure 6.2 presents a comparison of performance predictions, due to Cronje and Kentfield (1980), made using hand calculations (chain-dotted lines) and a numerical procedure as described in detail in Section 5.2 of Chapter 5. The initial pressure ratio was 4.0 and a strong entropy discontinuity

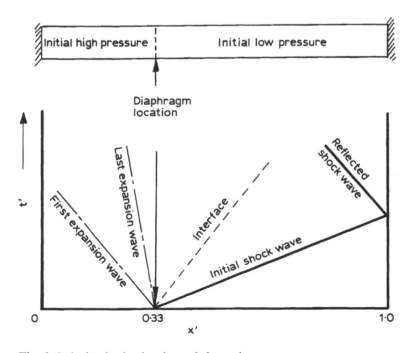

**Fig. 6–1** A simple shock tube and the major wave events.

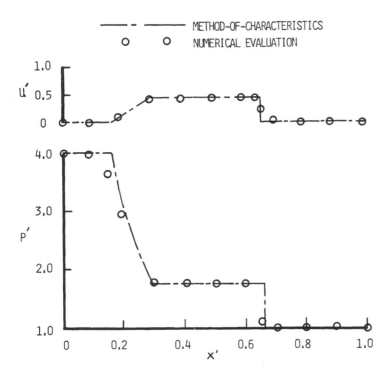

**Fig. 6–2** Predicted performance of the shock tube of Fig. 6.1 (initial pressure ratio 4, $\gamma = 1.4$, $t' = 0.245$).

prevailed between the high and low-pressure sections of the shock tube since the initial absolute temperature of the fluid in the high-pressure portion of the shock tube was assumed to be half that in the low-pressure portion. Figure 6.2 applies to dimensionless time $t' = 0.245$, based on the total length of the shock tube and the initial temperature of the low-pressure fluid, after rupture of the diaphragm.

Figure 6.3 shows similar information, for the same shock tube, $t' = 0.800$ after diaphragm rupture. The essential differences between the two diagrams relate to wave reflections having occurred at both the closed ends of the shock tube in Fig. 6.3 whereas, in Fig. 6.2, neither the expansion fan nor the initial shock had reached the closed ends. It can be seen from both Figs. 6.2 and 6.3 that apart from a tendency to "smear" discontinuities the accuracy of the numerical prediction is good. No account was taken of wall friction and heat transfer in either the hand or the numerical calculations, the results of which are presented in Figs. 6.2 and 6.3.

Figures 6.4, 6.5, and 6.6 show comparisons between experimentally obtained and analytical results, the latter based on the method-of-characteristics in

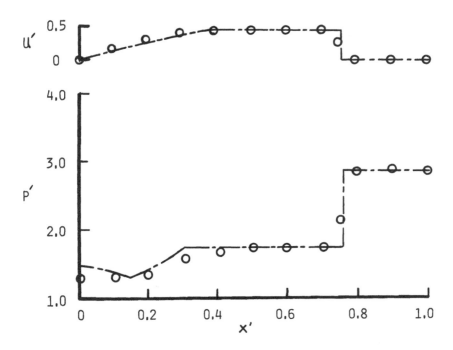

**Fig. 6–3** Predicted performance of the shock tube of Fig. 6.1 (initial pressure ratio 4, $\gamma = 1.4$, $t' = 0.800$).

conjunction with normal shock theory as applied in the previous example, reported by Glass (1958) for an air-filled shock tube having a large length-to-diameter ratio of approximately 400. The initial pressure ratio was 9.8 and the corresponding initial wall and charge temperatures were all equal. It might be expected that the differences between the analytical results (solid lines) and the experiments were due to the combined influences of wall friction and wall heat transfer, which were not taken into account in the theoretical analysis. This conclusion appears to bear credence since the superposition of numerically computed results in which wall heat transfer and friction were ignored (hollow symbols) compares quite well with the analytical results (solid lines), whereas when what happen to be suitable allowances were made for all friction and heat transfer the numerical model simulated, fairly closely, the results obtained experimentally. The numerically obtained results shown in Figs. 6.4, 6.5, and 6.6 have been presented previously (Cronje and Kentfield, 1980).

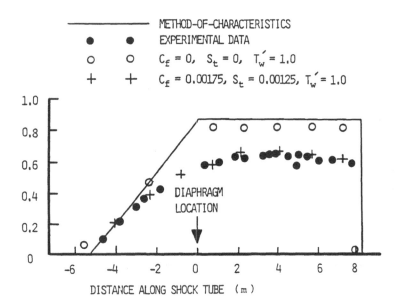

**Fig. 6–4** Predicted and experimentally measured velocity distributions in an air-filled shock tube 15 ms after diaphragm rupture. Initial pressure ratio 9.8.

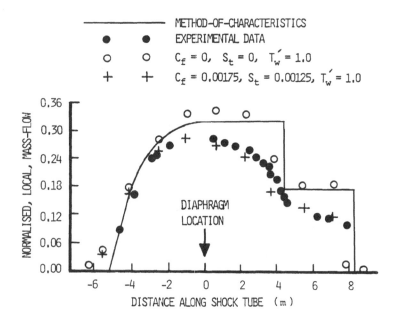

**Fig. 6–5** Predicted and experimentally measured normalized mass-flow distributions in the air-filled shock tube of Fig. 6.4 15 ms after diaphragm rupture (normalized with respect to initial conditions in the high-pressure region).

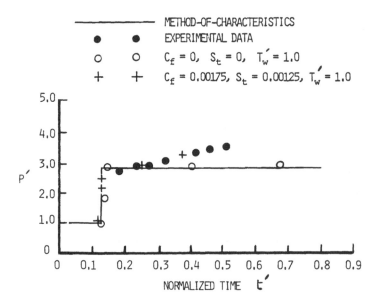

**Fig. 6–6** Predicted and experimentally measured pressure distributions, as functions of time, at a station 3 m downstream of the diaphragm location of the air-filled shock tube of Fig. 6.4.

## 6.2   Gas-Pipeline Flows

Transient flows in short, simple pipelines, for example a compressed-air pipeline in a small factory, while not usually of great practical interest can be dealt with using the techniques described previously in Chapters 2 to 5 inclusive. Two very elementary examples involving transient flows in pipes were included in Chapter 2, Section 2.3.

Usually of greater importance are transient flows in natural-gas distribution pipelines. Such pipelines often contain obstructions such as partially opened valves or what can be thought of as negative obstructions namely boost, or pressure make-up, compressors. Very often branches and junctions, the boundary conditions for which were considered in Section 4.5 of Chapter 4, are constituent components of natural-gas pipelines. Control valves that are regulated during pipeline operation must often be contended with. In common with most real pipelines bends or elbows are also virtually always an integral part of all natural-gas distribution systems.

Interest in the transient behavior of the flow field in natural-gas pipelines stems from the feasibility of allowing the pipeline to serve as a partial reservoir to assist in meeting either short-term customer overdemand at peak periods or the short-term accumulation of gas within the system when the rate of supply exceeds the customer demand. Such operation is desirable from the viewpoint of minimizing pipeline flow cross-sectional area, and hence capital cost, to meet a prescribed demand pattern. Such operation is, in large measure, rendered practical because, in most conventional systems, pressure regulators, or governors (i.e., automatically controlled pressure reducing valves) are installed at points where a main connects with a submain and where submains connect with subsub mains, etc. As a consequence, gas consumers are insulated from the influence of transients in the main supply system provided, of course, these are not of sufficient amplitude to overwhelm the capabilities of the pressure regulators.

Gas pipeline transients are here considered to be of a noncyclic nature and hence justify classification as initial value problems. The transients are assumed to be imposed on otherwise steadily flowing systems. Because of the very large length-to-diameter ratios, often in the region of $10^5$ or more, typical of natural-gas pipelines the influences of friction and heat transfer should always be taken into account when analyzing full systems. Hence a numerical analysis is virtually essential for obtaining solutions to complete problems although it is feasible to look at local events, for example in the region of changes of pipe cross section, etc., using simplified techniques in which friction and heat transfer influences are ignored. It is possible to establish, in this way, the underlying flow physics. It is rarely necessary to take into account potential energy terms when dealing with natural-gas pipelines since these only result in a maximum pressure, or density, reduction of about 5 percent relative to sea level values. A treatment of gas-pipeline flow, in which potential energy is taken into account, has been given by Wylie and Streeter (1978). Typically the normal steady-flow velocity in a gas pipeline is only about 10 m/s (30 ft/s). Hence in some problems the initial velocity prior to the imposition of transients can be regarded as negligible relative to the local acoustic velocity. The density of a typical natural gas is slightly greater than half that of air at the same pressure and temperature. The ratio, $\gamma$, of specific heats is in the region of 1.2.

### 6.2.1 Changes of cross-sectional area

The interaction of a compression, or expansion, wave with a transition in duct cross-sectional area results in the generation of both a transmitted and a reflected wave. Such an interaction is illustrated, diagrammatically, in Fig. 6.7 for a compression wave moving through a reduction, or contraction, of pipe cross-sectional area. It is assumed, for the simplified situation illustrated in Fig. 6.7 that the process is isentropic and that the dominant feature of the flow is the generation, from the incident compression wave, of the transmitted and reflected compression waves that propagate away from the contraction; in comparison with these the influence of minor irreversibilities within the contraction will be small. With reference to the state diagram, on the right-hand side of Fig. 6.7, the curved lines are lines of constant total pressure and take the form shown in Section 2.2 of Chapter 2:

$$a_0'^2 = a'^2 + \left(\frac{\gamma - 1}{2}\right)u'^2 \tag{6.1}$$

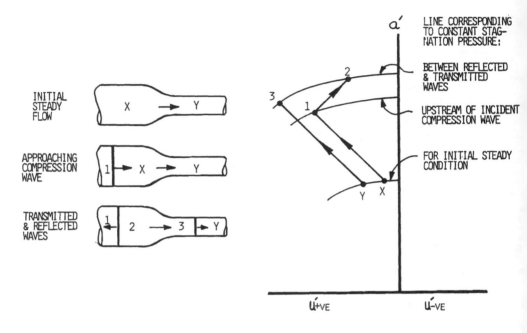

**Fig. 6–7** Rightward moving compression wave at a pipeline contraction: flow from left to right.

The state points $x$ and $y$ represent the initial flow conditions upstream and downstream, respectively, of the contraction. The corresponding final state points are 2 and 3, respectively. It is application of the continuity equations

$$\dot{m}_x = \dot{m}_y$$

and

$$\dot{m}_2 = \dot{m}_3$$

that leads to the results for isentropic flow

$$\left(\frac{a'_x}{a'_y}\right)^{2/(\gamma-1)} \cdot \frac{A_x}{A_y} = \frac{u'_y}{u'_x} \tag{6.2}$$

and, since $A_2 = A_x$ and $A_3 = A_y$,

$$\left(\frac{a'_2}{a'_3}\right)^{2/(\gamma-1)} \cdot \frac{A_x}{A_y} = \frac{u'_3}{u'_2} \tag{6.3}$$

Manipulation of Eqs. (6.1). (6.2), and (6.3) allows the final flow conditions to be established provided adequate information is known relating to the initial conditions and also the strength of the incident wave.

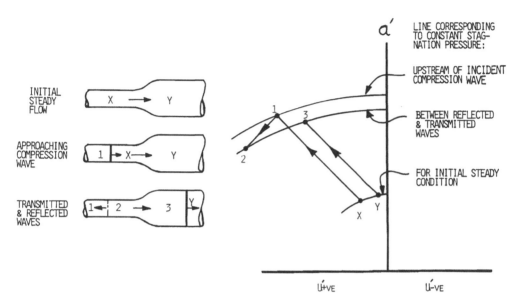

**Fig. 6–8** Rightward moving compression wave at a pipeline enlargement: flow from left to right.

The same analysis is applicable to an enlargement of the pipeline cross-sectional area. While it is likely that, in practice, the irreversibilities within an enlargement will be greater than for a contraction of equivalent area ratio, nevertheless, such losses will usually be small in comparison with those within the remainder of the pipeline system. The greatest influences of the enlargement will be the transmitted and reflected waves $y$–3 and 1–2, respectively, depicted in Fig. 6.8 that shows the events following the propagation of a rightward moving compression wave in fluid flowing from left to right through an enlargement. It can be seen from the state diagram that the reflected wave is of the expansion type. The transmitted wave is of the compression type but is weaker than the incident compression wave.

A rapid increase in demand at the downstream end of a pipeline will result in the propagation upstream of an expansion wave. The interaction of an upstream-propagating expansion wave with a pipeline contraction is depicted in Fig. 6.9 while the interaction of such a wave with an enlargement is shown in Fig. 6.10. In each case the notation selected is such that Eqs. (6.1), (6.2), and (6.3) can be applied to quantify the magnitudes of the transmitted and reflected waves.

### 6.2.2 Obstructions in pipelines

Localized obstructions in pipelines can take the form of, for example, venturi meters, partly open control valves or even a moving obstruction such as a

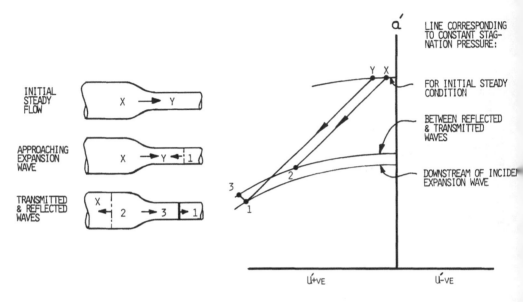

**Fig. 6–9** Leftward moving expansion wave at a contraction: flow from left to right.

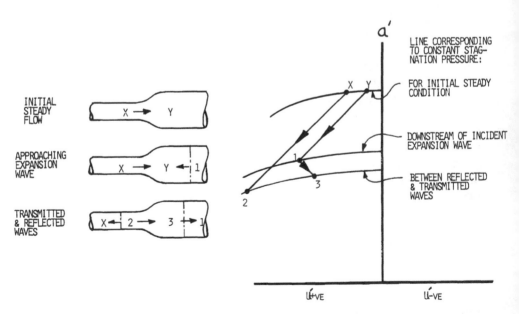

**Fig. 6–10** Leftward moving expansion wave at an enlargement: flow from left to right.

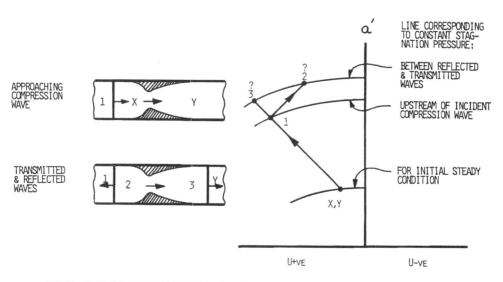

FOR A LOSS-FREE OBSTRUCTION POINTS 1, 2 & 3 MUST BE COINCIDENT HENCE WAVE Y-3 IS OF
EQUAL STRENGTH TO WAVE X-1 & CONSEQUENTLY WAVE 1-2 IS OF ZERO STRENGTH.

**Fig. 6–11** The transparency of a loss-free pipeline obstruction: flow from left to right.

pipeline pig. The latter is not, however, a form of obstruction expected to be
found under normal operating conditions. Long gas pipelines are usually
provided with intermediate compressor stations to boost the line pressure to
counteract the influence of line losses and hence to restore the gas density. Such
intermediate compressors can be treated as negative obstructions in a pipeline.

A venturi meter can be approximated as a loss-free obstruction since, for a
real venturi meter, the pressure loss occurring in practice is very small. In can
be assumed that a venturi will generate an upstream propagating reflected wave
in addition to a downstream transmitted wave much in the manner of, say, a
pipeline contraction. The treatment of a venturi meter in this manner is
illustrated in Fig. 6.11. State points $x$ and $y$ in Fig. 6.11 are coincident since
$A_x = A_y$ and the venturi is assumed to be loss free. After wave reflection and
transmittal the resultant state points can be expected to be as shown on the
state diagram constituting the right-hand portion of Fig. 6.11. However, states
2 and 3 must be coincident since the venturi is loss free and $A_2 = A_3$. The final
situation is, therefore, that state points 1, 2, and 3 are all coincident. Hence the
reflected wave 1–2 is of zero strength and the strength of the transmitted wave
$y$–3 is equal to that of the incident wave $x$–1. Thus the venturi meter, or any
other loss free obstruction, is in effect "transparent," the incident wave being
transmitted without reflection. The same result is obtained when the incident
wave is an expansion instead of the compression wave depicted in Fig. 6.11.

When a pipeline obstruction causes a significant loss of stagnation pressure,

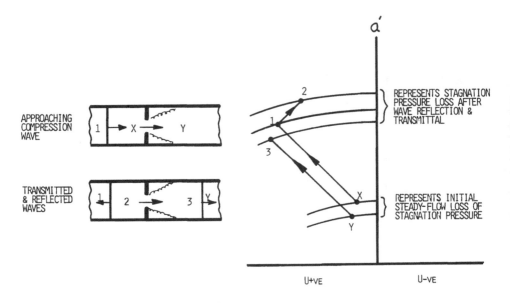

**Fig. 6–12** Transient flow through a nonloss-free obstruction such as a partly closed valve: flow from left to right.

for example, a partly closed control valve, the stagnation pressure of the flow downstream of the obstruction will be lower than that of the approaching flow. State points $x$ and $y$ of Fig. 6.12 represent the flows upstream and downstream, respectively, of a restriction across which there is a significant loss of stagnation pressure. A compression wave is shown, in the upper left-hand portion of Fig. 6.12, approaching the obstruction while the lower left-hand portion of the diagram shows the reflected and transmitted waves propagating away from the obstruction. The corresponding state diagram appears on the right of Fig. 6.12 although, in order to represent the events on the state diagram, the assumptions were made that an isentropic process connects state $x$ and state $y$ and, similarly, an isentrope connects the final states 2 and 3.

On the basis of the foregoing assumptions the strengths of the reflected and transmitted waves can be evaluated quantitatively by means of Eqs. (6.1), (6.2), and (6.3), for the special case $A_x = A_y$, provided the steady-flow pressure loss characteristics of the obstruction are also known. The pressure-loss characteristics of partly open control valves, etc., can normally be established from loss coefficient data as presented in appropriate handbooks. The loss of total pressure, as a fraction of the upstream static pressure, can be expressed, at least for small pressure drops in the form:

$$\frac{P_{ox} - P_{oy}}{P_x} = S \frac{\gamma}{2} M_x^2 \qquad (6.4)$$

where $S$ is the loss coefficient. The notation of Eq. (6.4) corresponds to that of Fig. 6.12. Equation (6.4) also applies, but with appropriately modified notation, for the flow situation prevailing after wave transmittal and reflection, i.e.,

$$\frac{P_{o2} - P_{o3}}{P_2} = S \frac{\gamma}{2} M_2^2$$

In addition to pressure loss data across the obstruction, it will also be necessary, of course, for sufficient information to be available to specify the initial flow conditions. A control valve that is actuated during operation of the pipeline can be assigned a quasi-steady loss coefficient, appropriate to the instantaneous setting of the valve, varying as a function of time.

The treatment of a pipeline intermediate pressure-boost compressor subjected to a line transient is illustrated in Fig. 6.13. In order to show the events on the $u' \sim a'$, state plane it has been assumed that the compressor compression process is isentropic. This assumption is an optimistic one for a turbocompressor and is pessimistic if the compressor is of the well-cooled reciprocating type. In the example depicted in Fig. 6.13 the incident wave is shown as a compression. Whether the reflected wave is a compression or an expansion depends, essentially, on the compressor operating characteristics and the mode of control employed. Both possibilities are indicated in Fig. 6.13, in each case the transmitted wave is a compression.

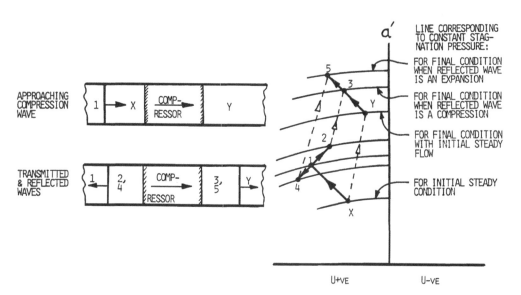

Fig. 6–13 Transient flow through a gas-pipeline compressor: flow from left to right.

### 6.2.3   Entry and exit valves

Valves, especially valves that are regulated as a function of time, at the entry and exit of a gas pipeline connected between large reservoirs, or the equivalent of large reservoirs, can be handled by means of the Jenny boundary relationships, as presented in Chapter 4. The Jenny boundary condition relations (Jenny, 1950) can also be applied quasi steadily when conditions within the large terminal reservoirs change, relatively slowly, with time.

### 6.2.4   Transient flow in bends

The analysis of transient flow around bends in ducts or pipelines is, at the very least, a problem in two space-dimensional nonsteady flow and, if account is also taken of secondary flow, attention should be paid to three, rather than two, spatial coordinates. For the present purposes attention will be given to the consideration of two space-dimensional nonsteady compressible flow.

Figure 6.14 shows the predicted propagation of an initially planar acoustic wave through stationary fluid within a 90° bend. The wave is traveling from left to right as it enters the bend at station 1. By the time the acoustic wave

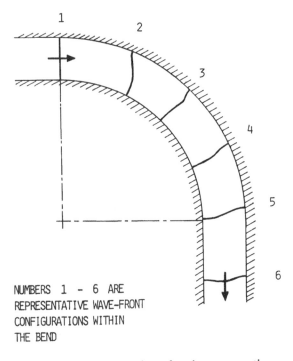

NUMBERS 1 - 6 ARE
REPRESENTATIVE WAVE-FRONT
CONFIGURATIONS WITHIN
THE BEND

**Fig. 6–14** The propagation of a planar acoustic wave through a two-dimensional right-angle bend.

has left the bend and arrived at station 6 it is very close to being planar never having been radically removed from the planar state as it propagated through the bend. Such a result should not be regarded as surprising since the operation of, for example, a speaking tube depends largely upon the ability of acoustic waves to propagate around bends.

The prediction presented in Fig. 6.14 was based on reflection, from the outer surface of the bend, of the incident acoustic wave, as indicated by Porges (1977), plus acoustic propagation into the stationary fluid within the bend from the "tail," or free end, of the wave. The evaluation was made graphically and should be regarded only as approximate.

When the incident acoustic wave is replaced by, say, a compression wave of finite strength, behind which the fluid is in motion and has a significant flow velocity towards the bend, the situation portrayed in Fig. 6.14 is modified. A two-dimensional compression wave is formed and propagated into the moving fluid on the outside of the bend raising the static pressure therein and, due to an upstream component of propagation, slowing the fluid approaching the bend. On the inside of the bend a two-dimensional expansion wave is formed that also has an upstream component of propagation. The two-dimensional expansion wave serves to both lower the static pressure adjacent to the inner surface of the bend and to increase the flow velocity of the fluid approaching the bend. The outer compression wave also increases the strength of the incident wave, and hence also increases the incident wave propagation velocity adjacent to the outside of the bend, whereas the inner expansion wave has the opposite effect adjacent to the inner surface of the bend. This situation is illustrated, diagrammatically, in the left-hand portion of Fig. 6.15. Both the outer compression wave and the inner expansion wave also perform the function of turning the flow so that it passes smoothly through the bend. Both the two-dimensional compression and the two-dimensional expansion waves attenuate in amplitude as they propagate away from their origins (Porges, 1977).

The raising of the static pressure toward the outside of the bend and the lowering of the static pressure toward the inside of the bend are conditions consistent with the well-known requirements for radial equilibrium for steady flow, namely:

$$\frac{\partial p}{\partial r} = \frac{\rho}{r} u_\theta^2$$

where $u_\theta$ is the peripheral velocity of flow in the bend. It can further be shown that, for a constant $\rho$, $u_\theta$ is inversely proportional to the radius $r$ when the flow field can be described as reversible, i.e., is consistent with the requirements of isentropic flow. The relatively low velocity prevailing adjacent to the outer surface of the bend in conjunction with the relatively high velocity adjacent to the inner surface are, therefore, conditions inherently consistent with the requirements for isentropic, or free vortex, flow within the bend.

The overall implication is, therefore, that a compression wave of finite strength can be expected to pass through a pipe, or duct, bend in such a way

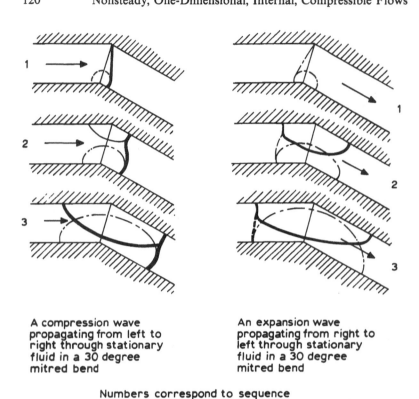

A compression wave
propagating from left to
right through stationary
fluid in a 30 degree
mitred bend

An expansion wave
propagating from right to
left through stationary
fluid in a 30 degree
mitred bend

Numbers correspond to sequence
(expansion-wave fanning ignored throughout)

**Fig. 6–15** The propagation of planar acoustic waves through a two-dimensional right-angle bend.

that, to a first approximation, the wave remains normal to the duct walls. The passage of the wave can also be expected to give rise to conditions consistent with a subsequent free-vortex type flow field within the bend. Similar implications relate to the propagation of expansion waves as indicated in the right-hand portion of Fig. 6.15 that shows an expansion wave propagating from right to left in a bend filled, initially, with stationary fluid.

It can be deduced, therefore, that provided a duct, or pipeline, bend is of adequate radius-to-flow-channel-width ratio pressure waves are transmitted essentially without inference and hence, with very little error, bends can be accounted for as simply equivalent lengths of straight duct or pipe. For bends with very small mean radii, i.e., 90° elbows or mitred right-angle bends, internal reflection is likely to occur with significant resultant attenuation. Attenuation due to this cause is particularly likely, as indicated by Porges (1977), when the incident event is of high frequency and has a wavelength approximately equal to the duct width or diameter. It would also appear to be possible to achieve a measure of attenuation, with a high frequency input signal, in a bend subdivided into a number of parallel channels, for example by turning vanes. Due to differences in flow-path lengths, the output of one channel may become phase-shifted relative to those of adjacent channels thereby resulting in a measure of attenuation in the region where the flows recombine downstream of the bend. High frequency disturbances are not, however, likely to be of prime concern in gas pipelines.

## Problems

NOTE. Assume, unless otherwise stated, that $\gamma = 1.4$.

6.1 An air-filled shock tube has an initial pressure ratio of 6:1. The frangible diaphragm is located half way between the two closed ends. Treating the initial compression wave in the low-pressure portion of the tube as a true normal shock, establish the pressure distribution in the tube at the instant the initial shock wave reaches the closed end. The contents of each portion of the shock tube are at room temperature prior to rupture of the diaphragm. Use as reference pressure, the initial pressure in the low-pressure portion of the tube and use the initial temperature of the shock tube contents as a reference temperature. What, in terms of dimensionless time based on the overall length of the shock tube and the reference temperature, is the time taken for the initial shock to reach the closed end? Use normal-shock tables, if desired, when establishing your solution. Ignore wall friction and heat transfer.

6.2 An infinite reservoir is connected to a pipe, length 3000 m, as illustrated in Fig. 6.16. The valve at station 1 is fully opened, for all practical purposes, instantaneously. Establish the pressure distribution in the tube 14 seconds

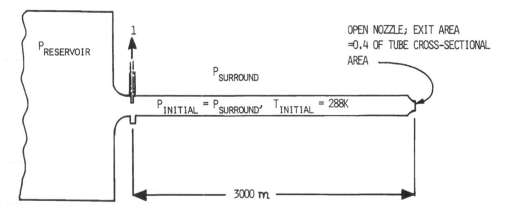

**Fig. 6–16** Diagram for problem 6.2.

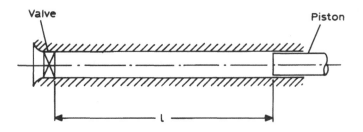

Valve　　　　　　　　　　　　　　　　　　　Piston

$l$

**Fig. 6–17** Diagram for problem 6.3.

after the valve is opened; the air in the tube initially is at rest and has the same pressure as the surroundings.

The ratio of the air pressure in the reservoir to the surroundings pressure is 1.74:1. The initial air charge in the tube has the same entropy as the contents of the reservoir.

Use the acoustic conditions in the pipe initially, and the length of the pipe, for reference purposes.

Establish, in addition, the flow Mach number in the pipe just upstream of the nozzle, and the Mach number in the nozzle exit, 14 seconds after the valve is opened.

6.3 A cylinder provided with a closely fitting piston is arranged as shown in Fig. 6.17. At time $t = 0$ the face of piston, which is at rest, is distance $l$ to the right of a quick-acting valve located at the left hand end of the cylinder.

At time $t = 0$ the valve opens instantaneously, and thereafter remains fully open, to expose the full cross-sectional area of the cylinder to the surroundings. The initial difference in pressure between the air in the cylinder and the surrounding air is such that flow enters the cylinder at a flow Mach number of 0.6. Use the acoustic velocity in the undisturbed surroundings for reference purposes and note that at the instant the initial wave reflects from the piston face the piston instantly commences to move toward the left-hand end of the cylinder at a steady, constant velocity of 20 percent of the reference acoustic velocity.

Establish, by means of the simplified isentropic treatment, the ratio of the:

　a. static pressure in the cylinder, at the instant the third wave reflected from the piston arrives at the open valve, divided by the initial cylinder pressure,

　b. static temperature in the cylinder, at the instant the third wave reflected from the piston arrives at the open valve, divided by the surroundings temperature.

Also sketch the wave diagram showing the Mach number, as a function of time, in the open end of the cylinder.

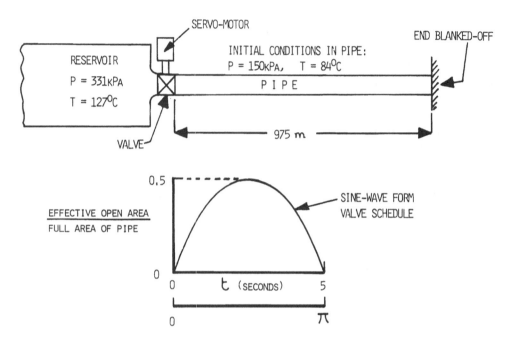

**Fig. 6–18** Diagram for problem 6.4.

Assume that there are no inflow pressure losses and that outflow occurs without pressure recovery.

6.4 A pipe, shown in Fig. 6.18, contains, initially, air at rest at a pressure of 150 kPa absolute (i.e., approximately 1.5 atmospheres) and at a temperature of 84°C. At time $t = 0$ the pipe is brought into communication with a reservoir containing air at a pressure and temperature of 331 kPa and 127°C, respectively.

   The communication between the pipe and the reservoir is regulated by means of a servocontrolled valve. The valve at first commences to open and subsequently closes in a total of five seconds as indicated. The effective valve port opening as a fraction of the full area of the pipe never exceeds 50 percent. Establish the pressure distribution in the pipe, normalized by the initial pipe pressure, at the instant the valve closes fully. The characteristic gas constant, $R$, for air $= 285$ Nm/kg·K.

6.5 Flow in an air filled pipeline is moving steadily through a contraction: the flow Mach number upstream of the contraction is 0.1. The ratio of the cross-sectional area of the pipe upstream of the contraction to that downstream is 2:1. A compression wave propagates, traveling in the same direction as the flow, toward the contraction. The ratio of the static pressure upstream of the compression wave to that downstream is 1.7. Establish the type, direction of travel, and static pressure ratio of the transmitted and reflected waves generated by the arrival, at the contraction, of the compression wave. Ignore irreversibilities within the contraction.

6.6 Air flowing steadily passes through a partly closed control valve in a pipeline. The flow Mach number is 0.1, in the pipe, upstream of the control valve. Under these conditions it is found that the control valve causes a reduction of static pressure equal to half of the absolute static pressure prevailing in the upstream pipe. A compression wave of static pressure ratio 1.7, propagating in the same direction as the flow, approaches the control valve. Establish the approximate pressure ratios of the subsequent transmitted and reflected pressure waves. Ignore pipe-wall friction and heat transfer.

# 7

## PRESSURE-EXCHANGERS

Pressure-exchangers—some workers use the alternative title energy-exchanger —is the generic name given to a class of machine that utilizes nonsteady compressible flow phenomena to perform a variety of useful, cyclic, processes. An essential feature of pressure-exchangers is a cellular component usually constituting the rotor of the machine. It is, perhaps, most convenient to describe the essential structure of a typical pressure-exchanger and, subsequently, to investigate the principles of the device. Equipped with this knowledge it is then possible to consider the potential advantages and disadvantages of pressure-exchangers and hence show why there is currently an increasing interest in machines of this type.

### 7.1 Structure, Principles, Advantages, and Disadvantages

A pressure-exchanger, in its most common form, comprises two basic components:

a. a rotor of cellular construction,

b. a stator containing gas ports in the end faces. The end plates of the stator are mounted as closely as possible to the rotor consistent with the avoidance of rubbing.

Figure 7.1 is a diagram of the basis of most pressure-exchangers. Only one port is shown in each end plate merely to illustrate the essential features of the device. Figure 7.1 does not show any means for locating the end plates in position. Usually a sleeve, passed over the rotor, is employed to connect the end plates together as indicated in Fig. 7.2.

In practice the outersleeve of Fig. 7.2 is often telescopic; this allows the clearances between the rotor and the end plates to be maintained at a constant, or very nearly so, predetermined value despite changes in rotor temperature, the end plates moving closer together, or further apart, as the rotor cools or heats up, respectively. A simpler alternative technique, less suited to transient situations, is to control the sleeve temperature such that correct clearances are maintained.

A means is provided to rotate the rotor; this may consist of an electric motor coupled to the rotor by, say, a belt transmission. Alternatively rotors can be made self-driving as will be explained later. In either case the power required

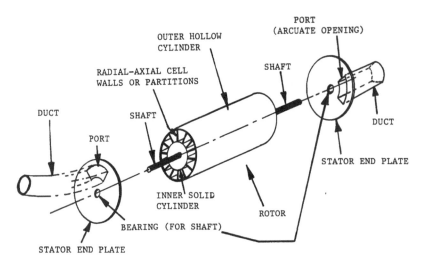

**Fig. 7–1** Basic arrangement of a simple pressure-exchanger (diagrammatic).

to maintain rotor rotation in a correctly designed pressure-exchanger is very small; it is nearly zero. The function of the driving means is merely to overcome bearing friction and rotor windage. There is always the inherent possibility in pressure-exchangers of rotating the "stator" and fixing the "rotor." This arrangement rarely seems to be convenient but it should, at least, be kept in mind.

So far only the hardware has been considered. Imagine, for example, that the rotor is spinning and that the cells contain gas at rest relative to the rotor.

If a port, for example the one on the left in Fig. 7.1, is supplied with fluid

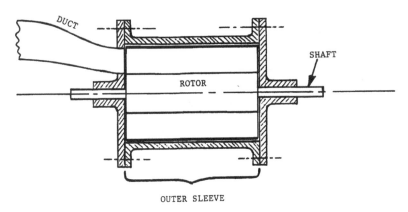

**Fig. 7–2** Outer sleeve of stator connecting the two end plates of a pressure-exchanger.

(gas) at a pressure higher than that in the cells then gas will enter each cell as it passes the port. The inflow will compress the contents of each cell in turn and will set it in motion to the right exactly as in a duct filling process. The rotor can be made self-driving by setting an inlet port at a sufficient incidence to the rotor, such that the inflow impinges on the cell walls.

Consider, as the alternative to cell filling, what happens when a pressure lower than that prevailing in the cells exists in a port. The port can be regarded as connected to a sink, or low pressure reservoir, by the duct coupled to it. Fluid leaves the cells via the port and passes to the reservoir. If, for example, the right-hand port of the basic pressure-exchanger of Fig. 7.1 is connected, via the duct, to a sink then outflow will occur from each cell as it passes the port. The outflow sets the contents of the cells in motion such that fluid flows out to the right, the consequent expansion lowering the pressure of the cell contents remaining. This is analogous to a duct emptying process.

It can be seen, in this way, that by cutting ports in the end plates and connecting these to appropriate high or low-pressure reservoirs fluid will enter, or leave, the rotor in a prearranged sequence. In most pressure-exchangers the frequency with which cells pass the ports is such that flow in the ports is (substantially) steady; the nonsteady flow being confined to the cells.

It remains to be shown that such an arrangement can be made to perform useful functions of practical interest. However, before proceeding further a number of general properties characteristic of pressure-exchangers can be identified. These can be classified as either an advantage or disadvantage of the concept.

*Advantages:*

   i. Robust construction.
  ii. Generally low rotational speed compared with turbomachines; too high a rotational speed will obviously not permit the occurrence of adequate cell filling or cell emptying.
 iii. Isentropic efficiencies of compression and expansion comparable with those of turbomachines; this follows if efficient cell filling and emptying processes are encouraged.
  iv. Less prone to erosion damage, due to solid particle or liquid droplet ingestion, than turbomachinery. The velocity of the working fluid in a pressure exchanger relative to the hardware is typically about one-third of values commonly met within turbomachines; this implies that the kinetic energy of a particle of prescribed mass is only about 10 percent, in a pressure-exchanger, of values attainable in typical turbomachines.
   v. Ability to withstand higher temperatures than say, gas turbines without recourse to structural cooling. The hottest parts of the rotor of a pressure-exchanger are usually only exposed to the maximum working temperature for a portion of the running time; in some cases

these regions are also in contact with cool working fluid such as, for example, scavenge air.

vi. Nonsurging performance characteristics of pressure-exchangers; pressure-exchangers do not surge in the manner customarily associated with axial or centrifugal turbocompressors.

*Disadvantages:*

i. Low mass flow rate per unit of frontal area compared with turbomachinery, particularly in relation to axial flow turbomachines.

ii. Not primarily suitable for the direct production of shaft power output without the use of a power turbine.

iii. Noise generation; due to the nonsteady flows occurring within them, pressure-exchangers are inherently noisy and therefore require careful muffling or silencing. The noise produced is predominantly of high frequency, usually say 1 to 5 kHz, and can, therefore, be dealt with fairly easily by absorption devices.

Before advancing to consider particular pressure-exchanger cycles it is worth noting that there are at least two classes of pressure-exchanger: dynamic pressure-exchangers and static pressure-exchangers. The dynamic pressure-exchanger, which makes use of pressure waves in the cells much as for the duct filling and emptying processes described in previous chapters, is currently, and is likely to remain, the most important class of pressure-exchanger.

### 7.2  Dynamic Pressure-Exchange

In dynamic pressure-exchangers, sometimes called pressure-wave-in-rotor devices that is usually abbreviated to pressure-wave devices or more simply wave rotors, the port width for a filling process is chosen to be such that the port trailing edge is passed by a cell at (ideally) the instant the compression wave reflected from the closed end arrives at the port. This situation is illustrated in Fig. 7.3.

It should be noted that in the notation used in Fig. 7.3, and in subsequent figures, compression waves are shown as solid lines, fluid demarcation (i.e., particle path) lines as dotted, and expansion waves as chain-dotted lines. This is consistent with the notation adopted previously for the general treatment.

It is normal convention to present unsteady flow fields on the $x \sim t$ plane; in the case of a pressure-exchanger the ordinate represents the elapse of time in a cycle from some arbitrary datum; the abscissa is directly proportional to distance along the channel (cell) in which the one-dimensional nonsteady flow occurs. In the case of the pressure-exchanger $x$ is measured along the axis of the cells. The arrow in the port represents the direction of flow; the arrow on the cells serves to remind the reader that the cells are, in view of the defintion of the time scale, running upward.

Figure 7.4 shows an emptying process in a dynamic pressure-exchanger.

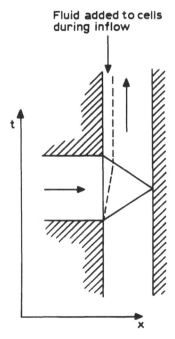

**Fig. 7–3** A basic cell filling process using dynamic pressure-exchange.

The expansion wave fanning which must, in reality, occur has been omitted for clarity.

There are some p.e. wave processes made up partly of compression and partly of expansion waves. Still others, called scavenging processes, are characterized by waveless regions. Examples of all these appear in Section 7.4.

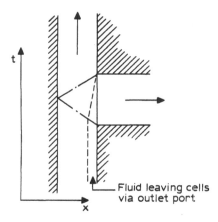

**Fig. 7–4** A basic cell emptying process using dynamic pressure-exchange.

The inherent difference between a dynamic pressure-exchange process and a so-called static process is dependent, essentially, upon the duration of port opening.

## 7.3  Static Pressure-Exchange

If the time taken for an acoustic wave (i.e., a wave traveling at sonic velocity) to travel the length of the cell and back again is very much less than the time taken for a cell to pass a filling (inlet) or an emptying (outlet) port then wave effects will die out and the pressure in the cells will equalize with that in the port. A pressure-exchanger in which this occurs is clearly unable to derive benefit from gas dynamic (pressure wave) effects and to distinguish it from pressure-exchangers that do utilize wave effects it is called a *static pressure-exchanger* or, sometimes, a *Lèbre* pressure-exchanger after its originator.

It is clear that since either throttling occurs during the uncovering, or covering-up, of the cell ends or, if wave effects are generated, they are not utilized, the ratio between the initial pressure in the cell and that in the port must be kept as close as possible to unity in order to minimize irreversibilities. A static pressure-exchanger is, therefore, characterized by numerous ports each at a slightly differing pressure to those adjacent to it. This situation implies a low rotor speed if each port is to be sufficiently narrow to permit many ports to be accommodated on each end plate.

Because of the relatively large number of ports involved, and the implication of low flow velocities resulting from the desirability of minimizing throttling losses, static pressure-exchangers are usually of less practical interest than those of the dynamic kind. However, it should be kept in mind that a dynamic pressure-exchanger operating at substantially less than the design rotor speed tends to function in the manner of a Lèbre-type machine.

## 7.4  Pressure-Exchanger Configurations

Dynamic pressure-exchangers can be configured to perform a wide variety of useful tasks. Six particular configurations are shown in Fig. 7.5 to Fig. 7.10 inclusive. In each case the ports of the pressure exchanger are identified by labeling them with reference to the function of each port in accordance with the following code:

L ≡ low pressure
M ≡ medium pressure     subscripts "IN" or "OUT" are employed to identify inlets or outlets, respectively
H ≡ high pressure

In order to simplify the diagrams each wave, compression or expansion, is identified as a single line. As before, solid lines are used for compression waves and chain-dotted lines for expansion waves. Particle paths are represented as dotted lines. Wave interactions at interfaces are ignored. For each cycle

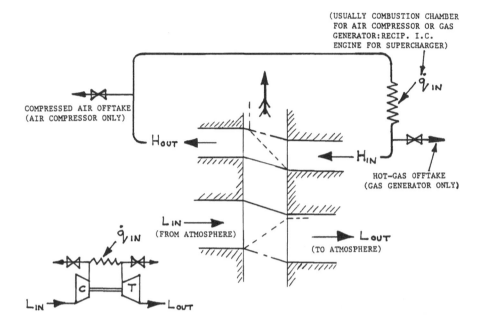

**Fig. 7–5** A gas-generator, air-compressor, or supercharger configuration utilizing dynamic pressure-exchange.

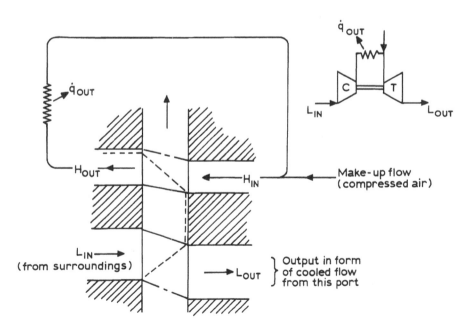

**Fig. 7–6** A dynamic-pressure-exchanger arranged as as air-cycle refrigerator.

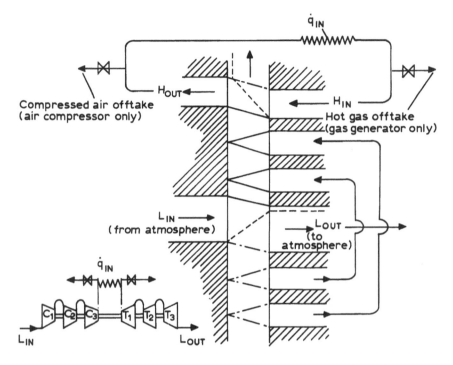

**Fig. 7–7** A gas-generator, air-compressor, or supercharger configuration incorporating two transfer passages.

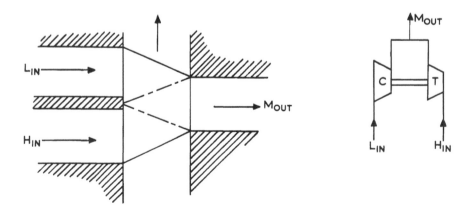

**Fig. 7–8** A dynamic pressure-exchanger equalizer.

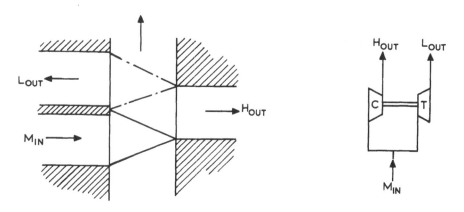

**Fig. 7–9** A dynamic-pressure-exchanger divider.

illustrated a turbomachine capable of performing the same task is shown, diagrammatically, alongside the dynamic pressure-exchanger that is drawn on the distance ∼ time plane. It should be borne in mind that, because the processes are cyclic, the top of each distance-time diagram can be considered to be looped around and joined to the bottom or, alternatively, each cycle is repetitive.

Figure 7.5 shows the basic configuration of a machine that can serve as a low pressure-ratio gas generator or air compressor or internal-combustion-engine supercharger. The cycle involves two scavenging processes—one at the low pressure of the cycle the other at the high-pressure condition. The

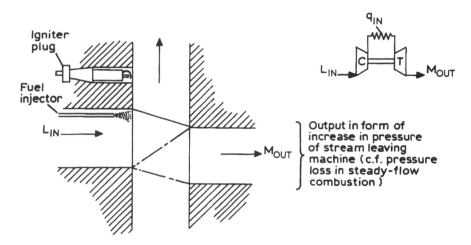

**Fig. 7–10** A pressure-gain combustor using dynamic pressure-exchange.

low-pressure scavenge process serves to replace products of combustion exhausting from the machine with an induced air flow from the surroundings. The high-pressure scavenge process results in discharging air, which has been compressed, into the high pressure circuit while receiving products of combustion from the same circuit. The cycle involves, in essence, two compression and two expansion waves.

A similar cycle, but with a heat rejection and mass injection replacing the heat addition process, is shown in Fig. 7.6. This machine can serve as an air-cycle refrigerator. The "output" flow is the cooled low-pressure stream leaving the unit.

Figure 7.7 is another configuration suitable for use as a gas generator, air compressor, or supercharger. The arrangement differs from that of Fig. 7.5 by virtue of the addition of transfer passages each connecting an emptying process to a filling process. The purpose of transfer passages is to increase the overall pressure ratio of the machine without recourse to very high gas velocities in the system. Two transfer passages are shown in Fig. 7.7: at least in principle there is no restriction on the number of transfer passages that can be used.

A pressure-exchanger equalizer is depicted in Fig. 7.8. This machine serves a similar function to an ejector or to the turbocompressor unit shown in the diagram. This cycle is interesting in that it is one which is wholly dependent on wave dynamics: there is no comparable cycle for a nondynamic, or static, pressure-exchanger. A similar remark can be made in relation to the pressure-exchanger divider cycle shown in Fig. 7.9. This cycle takes as an input a medium pressure stream and divides it into two outflowing streams, one at a higher and the other at a lower stagnation pressure than the input flow. There are potential uses for such a device in pressure-boosting applications.

Figure 7.10 illustrates a simple pressure-gain combustor concept. Here combustion of a fuel/air mixture takes place, at constant volume, within the cells of the pressure-exchanger. The resultant pressure rise results in the discharge of products of combustion at a higher stagnation pressure than the stagnation pressure of the entering air flow. The fuel flow can be added to the entering airstream in a number of ways.

The foregoing examples do not represent an exhaustive inventory of possible pressure-exchanger configurations; the applications are, however, sufficiently varied to give some idea of the wide range of uses to which pressure-exchangers can be put. It can be seen that dynamic pressure-exchangers represent applications of classical nonsteady one-dimensional compressible flow theory and hence the material already covered in Chapters 2, 3, 4, and 5 is directly applicable. The only additional factor to be considered is the influence of unavoidable leakage that occurs in the clearances, or gaps, between the stator end plates and the faces of the rotor. It is also worthwhile considering a simplified treatment of frictional effects. This allows an inviscid nonsteady flow-field analysis, such as one that can be carried out by hand, to be modified very simply to include, in an approximate way, the influences of frictional losses.

## 7.5   Influence of Friction

If it is desired to modify calculations carried out by means of the graphical method-of-characteristics to take frictional effects into account, a very simple, and rather approximate, procedure can be adopted. The justification for correcting wave diagrams, constructed without taking friction losses into account, is that in most practical problems associated with pressure-exchangers, frictional effects are relatively small.

The technique involves adding fictitious throttles to the inlet and outlet ports of a pressure-exchanger to take into account pressure losses due to:

- i. entry losses caused by cell wall blockage effects due to the finite thickness of the cell partitions,
- ii. exit losses also due to the finite thickness of the cell wall partitions,
- iii. cell wall friction.

The addition of the fictitious throttles to the ports of a pressure-exchanger can be pictured, notionally, as shown in Fig. 7.11 which depicts them, by way of example, in a divider. The wave events in a divider are shown in Fig. 7.9.

The procedures used to calculate the losses are based on standard steady-flow analysis concepts and are as follows:

i. *Entry losses:*

Defining $S_{ENTRY}$ as the multiple of the dynamic head in the entry port representing losses due to the sudden contraction at entry:

$$S_{ENTRY} = 0.5(1 - K_F) \qquad (7.1)$$

where:

$$K_F \equiv \frac{\text{flow area of cells exposed to the port}}{\text{flow area of the port}}$$

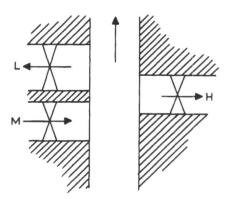

**Fig. 7–11** Illustration of fictitious throttles in the ports of a pressure-exchanger divider.

(*Note:* $K_F \to 1$ as $\Sigma$cell partition thickness $\to 0$.)

ii. *Exit losses:*

Similarly for outlet ports:

$$S_{\text{EXIT}} = (1 - K_F)^2 \tag{7.2}$$

Justification for the form of Eqs. (7.1) and (7.2) has been given by Spiers (1955). Similar justification can be found in most works on steady flow.

iii. *Friction losses:*

When dealing with pressure-exchangers it is convenient to base the loss due to friction to be assigned to each port as *half* that which would occur if fluid flowed steadily through the full length of the cells at the velocity and flow conditions prevailing in the port.

The factor of one-half was so chosen to take into account the average position of the pressure waves in the cells during a filling or emptying process. For other processes, for example scavenging, when flow takes place throughout the length of the cells the "half" rule means that half the losses are taken into account at the inlet end "throttle," the remainder at the outlet "throttle."

It can be shown, for example by Spiers (1955) and by many other authors, that in fairly smooth tubes such as those typically used for pressure-exchanger cells one dynamic head is lost in a length equal to approximately 60 hydraulic mean diameters. Thus:

$$S_{\text{CELL}} \simeq 0.016 l/d_h$$

where:

$S_{\text{CELL}}$    represents the number of dynamic heads lost over the *total* length of the cell

$d_h$    is the hydraulic mean diameter of the cells $= \dfrac{4 \times \text{cross-sectional area}}{\text{wetted perimeter}}$

Hence the total frictional losses to be assigned are as follows:

To inlet ports:

$$S_{\text{IN}} = S_{\text{ENTRY}} + (1/2)S_{\text{CELL}} \tag{7.3}$$

To outlet ports:

$$S_{\text{OUT}} = S_{\text{EXIT}} + (1/2)S_{\text{CELL}} \tag{7.4}$$

For a typical pressure-exchanger it was found that:

$$S_{\text{IN}} = 0.17, \qquad S_{\text{OUT}} = 0.12$$

Hence, it can be seen that the pressure losses assumed to occur in the port "throttles" are small for situations representative of dynamic pressure-exchangers. In these machines port Mach numbers rarely exceed (say) 0.6; the resultant maximum pressure loss at an inlet being less than 5 percent of the port stagnation pressure.

It should be remembered that a loss corresponding to $S_{IN}$ should be *added* to the pressure upstream of an inlet "throttle" and a loss corresponding to $S_{OUT}$ *subtracted* from the stagnation pressure downstream of an outlet "throttle."

A more elaborate treatment of viscous effects in a pressure-exchanger cell has been given by Cotter (1963).

### 7.6  Consideration of Leakage

So far nothing has been done to take leakage into account in either the computerized or noncomputerized method-of-characteristics procedures. Leakage is of main concern in the study of pressure-exchangers; in some other nonsteady flow situations its effects may be, and often are, absent. It is noteworthy that similar techniques can be used for both computerized and noncomputerized methods of analyzing pressure-exchanger performance.

Leakage in pressure-exchangers can be considered to be made up of two components:

i. circumferential leakage: i.e., from cell-to-cell or from cell-to-port within the clearance between the ends of the cell partitions and the stator and faces adjacent to them.

ii. radial leakage: i.e., from cells, via the inner and outer rims of the rotor, to the leakage sinks or vice-versa. Radial leakage can also occur from ports to the leakage sinks or the reverse.

Clearly the quantitative computation of leakage flows requires a knowledge of the clearances between the rotor and stator, the appropriate flow coefficients, and the pressure differential (pressure drop) across each leak in addition to knowledge of the relevant pressures and temperatures.

Because, in a well-designed pressure-exchanger, leakage effects, while not insignificant, are fairly small approximate methods can be employed to compute leakage flows and the effect of leakage on cycle performance.

Methods that have been proposed for computing leakage effects are:

i. linear leakage theory in which leakage in the vicinity of the ports is not taken into account,

ii. partly open end method; in this technique the method-of-characteristics calculations are carried out as if the cell ends never quite close, i.e., $\Phi_{MIN} > 0$.

iii. leakage model method; a leakage model is employed in which individual leaks are treated in such a way that leakage calculations can be superimposed upon the results of characteristics calculations carried out without taking leakage into account. These techniques will be examined in more detail.

### 7.6.1   Linear leakage theory

In this theory, due to Spalding (1956), radial leakage from the cells to the leakage sinks also the circumferential leakage from cell-to-cell in a direction from the high-pressure to the low-pressure port regions of the pressure-exchanger cycle are both taken into account in such a manner that the effect of leakage on the pressures in the cells is also included. The theory does not take into account the effect of wave action within the pressure-exchanger on leakage; also implicit assumptions are that the number of cells is very large and that the ports only occupy a small fraction of the stator end faces and, hence, all the important leakage occurs from cell-to-cell and from the cells to the leakage sinks.

The linear leakage theory, so named after the type of differential equation describing the leakage, is therefore approximately correct for pressure-exchangers in which the ports cover only a small fraction of the stator end faces. Pressure-exchangers of this type are not common although some pressure-exchangers made especially for research purposes fit this description. However the majority of pressure-exchangers tend to have stator end faces fitted with as many ports as possible in order to maximize the mass flows.

### 7.6.2   Partly open-end method

In this technique, in which it is (in effect) assumed that the minimum value of $\Phi$, the proportion of cell opening, is greater than zero, flow is considered to pass into or out of the "partly open end" even when a cell is not aligned with a port. The minimum value of $\Phi$, is, in principle, based on the clearances between the rotor and the stator and the appropriate flow coefficients. In practice, $\Phi$ is likely to be a variable; this will be the result of apportioning the geometrically derived value of $\Phi$ between the four component leaks at each end of each cell. The reason for this is that each component of the leakage may be either an inflow, an outflow or zero; $\Phi_{\text{MIN(EFFECTIVE)}}$ then being the appropriate effective opening of the cell end, either as an inlet or an outlet, that would permit the net flow into, or out of, each cell to be handled as for conventional partly open situations as described in Chapter 4, integrally, in the method-of-characteristics calculations.

The four components of leakage, and some possible leakage directions, are shown in Fig. 7.12. The partly open-end method, as outlined here, is probably best suited to the computerized method-of-characteristic, although it is applicable in principle to hand computations.

### 7.6.3   Leakage-model method

In this technique the approximation is made that the leakage in a pressure-exchanger can be approximated as a series of flows through well-defined channels in the manner shown in Fig. 7.13 for the leakage model of a filling process. The upper and lower dashed lines in Fig. 7.13 mark the boundaries of the region from (and to) which leakage is assumed to occur. The dotted boundary in the right-hand portion of Fig. 7.13 represents the boundary of the

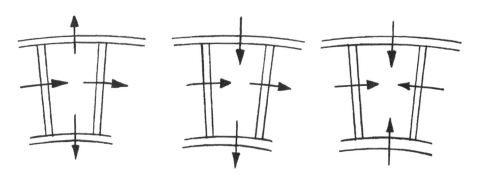

**Fig. 7–12** Components of leakage and possible leakage directions.

region covered by the filling process wave diagram or computerized method-of-characteristics computations: it should be noted that none of the leakage passages in the leakage model cross this dotted boundary.

The leakage computations for the filling process illustrated consist of applications of the First Law of Thermodynamics to the (notational) narrow regions *A* and *B* situated between adjacent wave processes and within which

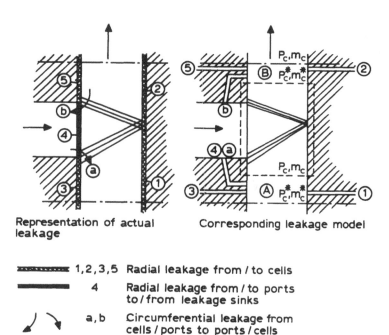

**Fig. 7–13** Leakage paths and corresponding leakage model.

the leakage effects of the filling process are assumed to be concentrated. Thus applying the First Law of Thermodynamics to region $A$:

$$\dot{N}(m_c T_c - m_c^* T_c^*)_A C_v = [C_p \Sigma \dot{m}_l T_l]_A \tag{7.5}$$

where:

$\dot{m}_l \equiv$ leakage mass flow ($+ve$ into cells)
$\dot{N} \equiv$ number of cells passing a fixed point per unit time.

Substituting as follows in (7.5)

$$m_c = \frac{P_c A_{\text{CELL}} l}{R T_c}, \qquad m_c^* = \frac{P_c^* A_{\text{CELL}} l}{R T_c^*}$$

$$\dot{N} = \frac{1}{\Delta t_{\text{CELL}}} = \frac{a_{\text{REF}}}{\Delta t'_{\text{CELL}} l}$$

the following result is obtained:

$$\left[1 - \frac{P_c^*}{P_c}\right]_A = \left[\frac{R \Delta t'_{\text{CELL}} \gamma}{P_c A_{\text{CELL}} a_{\text{REF}}} \Sigma \dot{m}_l T_l\right]_A \tag{7.6}$$

An exactly comparable expression, namely:

$$\left[1 - \frac{P_c^*}{P_c}\right]_B = \left[\frac{R \Delta t'_{\text{CELL}} \gamma}{P_c A_{\text{CELL}} a_{\text{REF}}} \Sigma \dot{m}_l T_l\right]_B \tag{7.7}$$

is obtained for region $B$.

It has been shown by Kearton and Keh (1952) that for simple leaks, that is leaks of the nonlabyrinth type, that up to and including choking:

$$\dot{m}_l = A_l C_S \sqrt{\frac{2\gamma}{\gamma - 1} \frac{P_l^2}{R T_l} \left[\left(\frac{P_x}{P_l}\right)^{2/\gamma} - \left(\frac{P_x}{P_l}\right)^{(\gamma+1)/\gamma}\right]} \tag{7.8}$$

and that for leakage past a labyrinth:

$$\dot{m}_l = A_l C_L \sqrt{\frac{F_K(P_l^2 - P_{n_L}^2)}{n_L R T_l}} \tag{7.9}$$

where:

$P_x$ = static pressure at the discharge of a simple leak,
$P_l$ = (effective) stagnation pressure at entry to leak,
$P_{n_L}$ = static pressure downstream of last ($n_L$th) labyrinth in a labyrinth seal of $n_L$ stages,
$F_K$ = function of $P_l P_L$ (displayed in Fig. 14 of Kearton and Keh's paper),
$n_L$ = number of stages in a labyrinth seal,
$A_l$ = cross-sectional area of leakage path,
$C_S$ = discharge coefficient of a simple leak,
$C_L$ = mean discharge coefficient of a labyrinth seal.

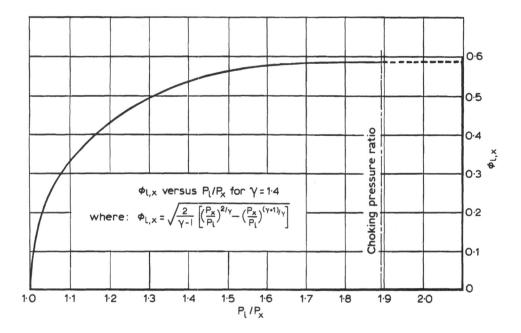

**Fig. 7–14** Leakage function $\Phi_{l,x}$ versus leakage pressure ratio $P_l/P_x$.

Substitution of expressions (7.8) and (7.9) into Eqs. (7.6) and (7.7) allows $P_c^*/P_c$ to be evaluated for regions $A$ and $B$, respectively, provided (of course) the areas and flow coefficients of the various leakage paths can be estimated reasonably accurately. Fortunately, as previously stated, leakage effects are relatively small in a well-designed pressure-exchanger and hence great accuracy is not, in most cases, essential.

A curve showing $\Phi_{l,x}$ versus $P_l/P_x$ where:

$$\Phi_{l,x} \equiv \sqrt{\frac{2}{\gamma - 1}\left[\left(\frac{P_x}{P_l}\right)^{2/\gamma} - \left(\frac{P_x}{P_l}\right)^{(\gamma + 1)/\gamma}\right]}$$

is presented in Fig. 7.14. This reduces the labor required to evaluate leakage flows by hand for simple leaks, the type most commonly met with.

It can be seen that the effects of including leakage in the manner shown here are:

i. to modify the matching condition between adjacent wave processes compared with that prevailing when leakage is absent,

ii. to modify the port mass flows and Mach numbers, etc.

It is also possible to apply the leakage model method to cyclic (as distinct from individual) wave processes. Examples of this type have been given by Kentfield (1963).

### 7.6.4  Leakage effects in practical situations

The significance of leakage effects in actual pressure-exchangers will be identified in more detail later.

It has, in general, been found to be possible to maintain sufficiently leak free operation for most purposes by simply employing fairly small clearances. These are typically in the region of a one thousandth part of the axial, and diametrical, dimensions of the rotor. Thus a pressure-exchanger with a rotor of, say, 200 mm length and 120 mm diameter would have a total axial clearance of 0.2 mm, or 0.1 mm between each end face of the rotor and the adjacent stator end plate, and 0.12 mm diametrical clearance, or 0.06 mm radial clearance.

## 7.7  Approximate Method for Performance Prediction

So far two theoretical procedures have been introduced—one a very simple technique merely for predicting idealized isentropic performance characteristics of pressure-exchangers and other nonsteady flow systems; the other a very much more sophisticated procedure, namely, the method-of-characteristics, capable of predicting the performance of pressure-exchangers (and other nonsteady flow devices) in considerable detail including, if required, heat transfer and frictional effects. The approximate procedure described here makes it possible, at the price of some loss of accuracy, to predict the performance of pressure-exchangers (only) nearly as comprehensively as with the method-of-characteristics by means of a technique not very much more complicated than that of the idealized prediction procedure of Chapter 2. The approximate procedure takes the effects of leakage and friction into account implicitly; a disadvantage is that it does not provide the wealth of detailed information, relating to flow within the cells, obtainable from the method-of-characteristics.

The approximate method, which is based on a procedure described previously by the author (Kentfield, 1968), is therefore of most use when making rapid exploratory feasibility investigations and when carrying out preliminary design studies, etc.

### 7.7.1  Outline of method

The simplified procedure is based upon the generalization of experimentally obtained cell filling and emptying data. The generalization of the filling and emptying data is carried out in such a way that any pressure-exchanger cycle can be synthesized and its performance estimated accordingly. The heart of the procedure is a four step process. The first three steps are carried out once and for all for each set of cell filling and emptying data and only need to be repeated when new, or revised, input data become available. The details of the fourth step depend upon the particular process or cycle being synthesized.

In addition to the four step main procedure, two auxiliary steps will be described. One is used only in the analysis of scavenging processes; the second auxiliary step allows the approximate port widths and relative phase settings to be deduced corresponding to the synthesized performance characteristics.

*The four step procedure*

The first four steps can be outlined as follows:

1. The development of a generalized form of energy equation applicable to simplified models of pressure-exchange processes involving (nominally) only a single pressure wave.

2. Determination from experimental cell filling and cell emptying data, used in conjunction with the generalized form of energy equation, of *static* pressure ratio ($\sigma$) versus *static* temperature ratio ($\tau$) relations for compression and expansion waves.

3. Establishment, with the aid of the generalized energy equation and $\sigma$ versus $\tau$ relations, of Mach number versus static pressure ratio curve families for compression and expansion wave processes.

4. Pressure exchange cycle calculations performed with the assistance of the $\sigma \sim \tau$ curves of step (2) and the $M \sim \sigma$ curve families of step (3).

   *Note:* If scavenge type processes occur if will also be necessary to introduce the first of the auxiliary steps previously referred to; the second auxiliary step will also be required if it is desired to obtain predictions of the port widths and phasings.

*Assumptions*

The fundamental assumptions applying to the simplified models of single wave processes, to which the generalized form of the energy equation is applied, are:

a. fluid states and velocities throughout each cell are taken as uniform at the commencement and completion of each single wave process,

b. there is no composition or temperature discontinuity at the interface between fluid entering during a filling process and the original contents of the cells,

c. velocity distribution assumed to be uniform across each port.

The following assumptions are implied as a consequence of the three fundamental assumptions:

d. end effects due to wave reflections at boundaries are absent; this is a consequence of assumption (a).

e. pressure waves are infinitesimally thin in the vicinity of boundaries. This assumption arises from the consideration that the flow velocities adjacent to solid boundaries must be zero in directions normal to these boundaries, or the normal velocity must be uniform (c) in the ports and cells when waves overlap ports. Since end effects (d) are ruled out neither of these situations are compatible with gradual changes of normal velocity across a thick wave.

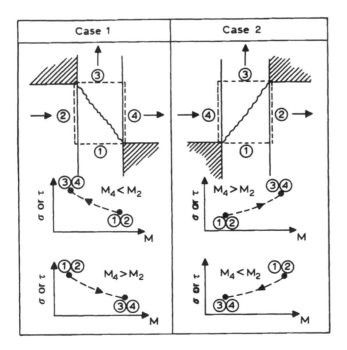

**Fig. 7–15** Illustration of the two basic wave processes.

It should be noted that the foregoing assumptions do *not* prevent account being taken of expansion wave fanning, the finite time taken for a cell to be opened or closed and so on, occurring in real processes. These effects are included automatically in the correlation of the experimentally obtained filling and emptying data with the simplified model processes. The latter are illustrated in Fig. 7.15.

### 7.7.2  Detailed description

Each of the four steps of the method will now be considered in more detail.

### Step 1.  Generalized Energy Equation

Application of the energy and continuity equations to the flow crossing the boundaries of either of the regions (Case 1 or Case 2) enclosed by dotted lines in Fig. 7.15 leads to a common result provided the previously stated assumptions are taken into account. The following derivation is in a form directly applicable to Case 1.

Equation (7.10) is obtainable by writing an energy equation for the system enclosed by the dotted boundary bearing in mind that fluid at states 1 and 3 is carried across the dotted boundaries in a rigid assembly of cells of constant volume, thus after canceling the whirl component of kinetic energy terms from both sides of the equation:

$$\dot{q} + c_v \dot{m}_1 T_1 + \dot{m}_1 \frac{u_1^2}{2} + c_p \dot{m}_2 T_2 + \dot{m}_2 \frac{u_2^2}{2} = c_p \dot{m}_4 T_4 + \dot{m}_4 \frac{u_4^2}{2} + c_v \dot{m}_3 T_3 + \dot{m}_3 \frac{u_3^2}{2}$$

(7.10)

where:

$\dot{q}$ = rate of heat flow from the structure of the p.e. into the region enclosed by the dotted boundary.

From the previously given assumptions for the regions enclosed by the dotted boundaries:

$$P_1 = P_2, \qquad u_1 = u_2, \qquad T_1 = T_2$$
$$P_3 = P_4, \qquad u_3 = u_4, \qquad T_3 = T_4$$

(7.11)

Applying continuity to flow across the dotted boundary:

$$\dot{m}_1 + \dot{m}_2 = \dot{m}_3 + \dot{m}_4$$

(7.12)

The following result is obtained after dividing Eq. (7.10) by $C_v \dot{m}_1 T_1$ and invoking Eqs. (7.11) and (7.12) and making some algebraic manipulations:

$$\frac{(\gamma - 1)\left[1 - \dfrac{\dot{m}_3 T_4}{\dot{m}_1 T_2}\right] - \dot{Q}}{\gamma\left[1 + \dfrac{u_2^2}{2\gamma c_v T_2} - \dfrac{T_4}{T_2} - \dfrac{u_4^2}{2\gamma c_v T_2}\right]} = \left[1 + \frac{\dot{m}_2}{\dot{m}_1}\right]$$

(7.13)

where:

$$\dot{Q} \equiv \dot{q}/c_v \dot{m}_1 T_1$$

The following equation can be written, using the steady-flow expression for continuity, bearing in mind that, in conformity with assumption (e), the pressure wave is taken to be infinitesimally thin (therefore $A_2 = A_4$) hence

$$\dot{m}_4 = \dot{m}_2 \frac{P_4 T_2 u_4}{P_2 T_4 u_2}$$

(7.14)

The following equation is obtained by substituting from (7.14) for $\dot{m}_4$ in (7.12), dividing by $\dot{m}_1$, and rearranging:

$$\frac{\dot{m}_2}{\dot{m}_1} = \frac{\left[\dfrac{\dot{m}_3}{\dot{m}_1} - 1\right]}{\left[1 - \dfrac{P_4 T_2 u_4}{P_2 T_4 u_2}\right]}$$

(7.15)

Because the volume of each cell remains constant:

$$\frac{\dot{m}_3}{\dot{m}_1} = \frac{P_3 T_1}{P_1 T_3}$$

(7.16)

It may also be shown that

$$\frac{u_2^2}{2C_v T_2} = \gamma\left(\frac{\gamma - 1}{2}\right)M_2^2 \tag{7.17}$$

and

$$\frac{u_4^2}{2C_v T_2} = \frac{T_4}{T_2}\gamma\left(\frac{\gamma - 1}{2}\right)M_4^2$$

The following result is obtained by making substitutions from Eqs. (7.15) to (7.18) inclusive in Eq. (7.13). It is the required generalized form of energy equation applicable to the model single wave process enclosed within the dotted boundary:

$$\frac{(\gamma - 1)\left[1 - \dfrac{P_4}{P_2}\right] - \dot{Q}}{\gamma\left[\left(1 + \dfrac{\gamma - 1}{2}M_2^2\right) - \left(1 + \dfrac{\gamma - 1}{2}M_4^2\right)\dfrac{T_4}{T_2}\right]} = \frac{\dfrac{P_4 T_2}{P_2 T_4} - 1}{\left[1 - \dfrac{P_4}{P_2}\sqrt{\dfrac{T_2}{T_4}}\dfrac{M_4}{M_2}\right]} + 1 \tag{7.19}$$

Equations (7.20) and (7.21) follow (after some algebra) for the special cases in which $M_2$ and $M_4$ are each made equal to zero, respectively.

When $M_2 = 0$ Eq. (7.19) reduces to

$$1 = \gamma\left[1 + \frac{\gamma - 1}{2}M_4^2\right]\frac{T_4}{T_2} - (\gamma - 1)\frac{P_4}{P_2} - \dot{Q} \tag{7.20}$$

and when $M_4 = 0$ Eq. (7.19) reduces to

$$\frac{P_4 T_2}{P_2 T_4}\left[\gamma\left(\frac{\gamma - 1}{2}\right)M_2^2 + \gamma - \frac{T_4}{T_2}\right] = (\gamma - 1) - \dot{Q} \tag{7.21}$$

Equations (7.19) to (7.21) inclusive can also be derived from a similar analysis for Case 2. Equation (7.20) applies to the first parts of both simple cell emptying (Case 1) and simple cell filling (Case 2). Equation (7.21) is applicable to the latter parts of both simple cell filling (Case 1) and simple cell emptying (Case 2).

### Step 2. $\sigma \sim \tau$ Relations

Adiabatic forms ($\dot{Q} = 0$) of Eqs. (7.20) and (7.21) were solved for temperature ratio ($T_2/T_4$ or $T_4/T_2$, general symbol $\tau$) using appropriate inputs of cell-to-port static pressure ratio ($P_2/P_4$ or $P_4/P_2$, general symbol $\sigma$), and port Mach number ($M_2$ or $M_4$), obtained from the results of cell filling and cell emptying experiments.

The $\sigma \sim \tau$ relationships established by this means are presented in Fig. 7.16 for cells of dimensionless width $\Delta t'_{\text{CELL}} \simeq 0.54$. A cell width of $\Delta t'_{\text{CELL}} \simeq 0.54$ was the only one for which filling and emptying data were available. The test results contributing to Fig. 7.16 were for tests with the port width adjusted to yield optimum filling, or emptying, performance with cells of the specified size, using air as working fluid.

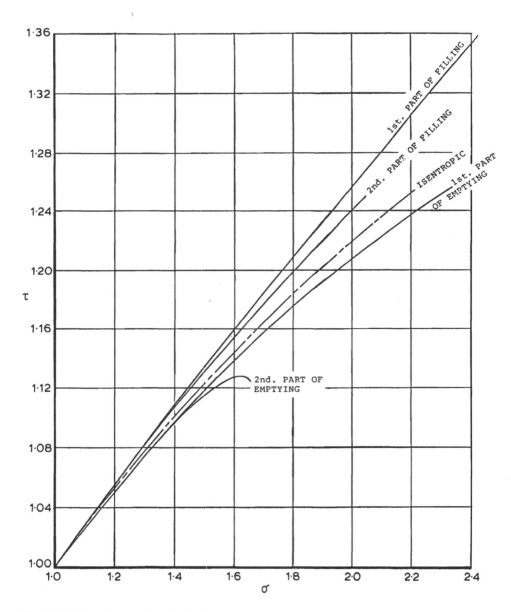

**Fig. 7–16** Experimentally obtained static temperature ratio ($\tau$) versus static pressure ratio ($\sigma$) relationships ($\Delta t'_{\text{CELL}} \simeq 0.54$, $\gamma = 1.4$).

*Step 3. M ~ σ Relations*

The curve families shown in Figs. 7.17 to 7.20 inclusive were constructed from evaluation (with $\dot{Q} = 0$) of Eq. (7.19) using the appropriate $\sigma \sim \tau$ relationship from Fig. 7.16. The choice of the particular $\sigma \sim \tau$ relation used in the evaluation of Eq. (7.19) was governed by the nature of the wave process under consideration; whether it was akin to the first or second part of a filling, or the first or second part of an emptying process. The $\sigma \sim \tau$ relation used in its construction, and the process to which it applies, are written on each curve family.

Curves (Figs. 7.17 to 7.20) inclusive constitute the basic tools needed for cycle calculations, within the limitations of the previously stated assumptions (a) to (e), when air is used as the working fluid (and when scavenging processes are absent and also when knowledge of port widths is not required) for $\Delta t_{CELL} \simeq 0.54$.

*Step 4. Application*

Application of the method can best be studied by means of examples.

*7.7.3   Illustrative examples*

*Example A*

A cell filling process takes place with a port Mach number of 0.3. What is the ratio of:

  a. the initial cell pressure to the port static pressure?

  b. the final cell pressure to the port static pressure?

(The cell width, $\Delta t'_{CELL} \simeq 0.54$, $\gamma = 1.4$.)

  a. From Fig. 7.18 (first part of filling) when $M_4 = 0.3$ ($M_2 = 0$), $P_4/P_2 = 1780$, or $P_2/P_4 = 0.562$, i.e.,

$$\frac{P_{CELL\,INITIAL}}{P_{PORT\,STATIC}} = 0.562$$

  b. From Fig. 7.17 (second part of filling) when $M_2 = 0.3$ ($M_4 = 0$), $P_4/P_2 = 1.430$, i.e.,

$$\frac{P_{CELL\,FINAL}}{P_{PORT\,STATIC}} = 1.430$$

*Example B*

In a divider (Fig. 7.9), in which $\Delta t'_{CELL} \simeq 0.54$, the medium pressure fluid enters at a Mach number of 0.5; the ratio of the static pressure in the HP outlet divided by that in the MP inlet = 1.2. What is the flow Mach number in the HP port? $\gamma = 1.4$.

From Fig. 7.17 when $M_2 = 0.5$ and $P_4/P_2 = 1.2$, then $M_4 = 0.347$, i.e., $M_H \simeq 0.35$.

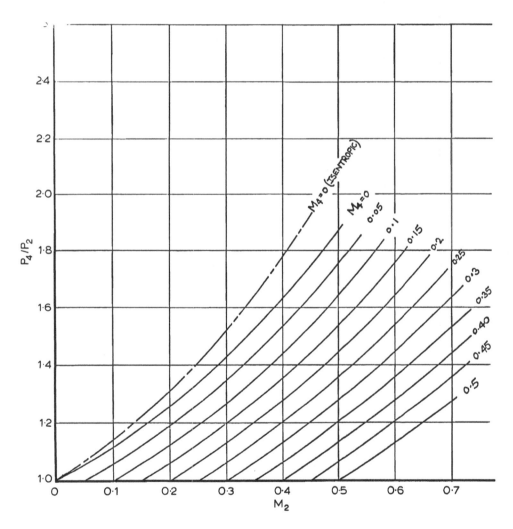

**Fig. 7–17** $P_4/P_2$ for Case 1, 2nd part of filling ($\Delta t'_{\text{CELL}} \simeq 0.54$, $\gamma = 1.4$).

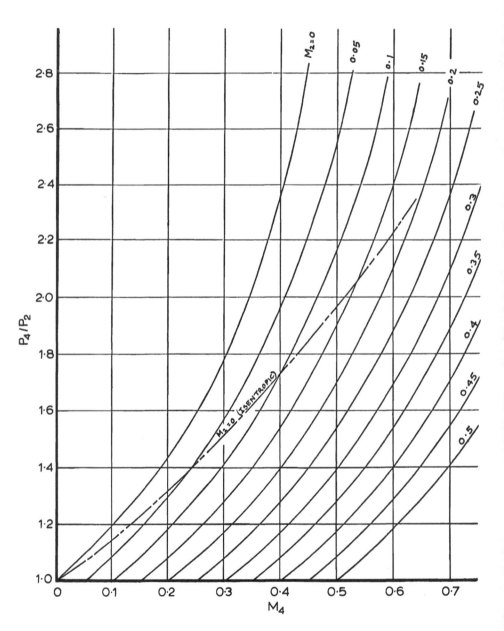

**Fig. 7–18** $P_4/P_2$ for Case 2, 1st part of filling ($\Delta t'_{\text{CELL}} \simeq 0.54$, $\gamma = 1.4$).

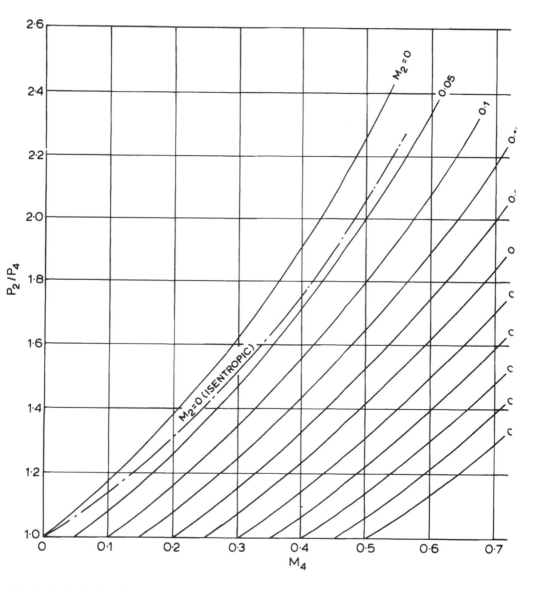

**Fig. 7–19** $P_2/P_4$ for Case 1, 1st part of emptying ($\Delta t'_{\mathrm{CELL}} \simeq 0.54$, $\gamma = 1.4$).

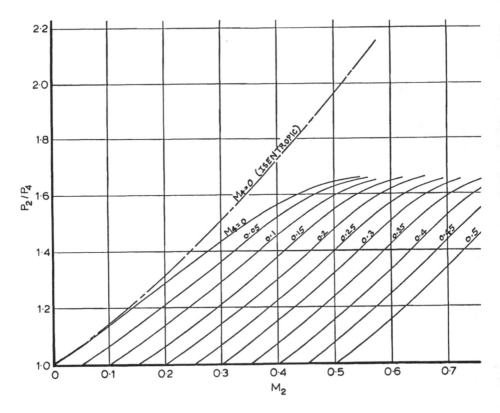

**Fig. 7–20** $P_2/P_4$ for Case 2, 2nd part of emptying ($\Delta t'_{CELL} \simeq 0.54$, $\gamma = 1.4$).

*Example C*

A cell emptying process takes place in which the ratio of the initial cell pressure to the port static pressure is 2.0. The cell width is such that $\Delta t'_{CELL} \simeq 0.54$, $\gamma = 1.4$.

    a. What is the outflow Mach number

    b. What is the final pressure within the cell as a fraction of the port static pressure?

    a. From Fig. 7.19, when $P_2/P_4 = 2.0$ and $M_2 = 0$, then $M_4 = 0.423$, i.e., $M_{OUT} = 0.423$.

    b. From Fig. 7.20, when $M_2 = 0.423$ and $M_4 = 0$, then $P_2/P_4 = 1.590$, or $P_4/P_2 = 0.629$, i.e.,

$$\frac{P_{CELL\ FINAL}}{P_{PORT\ STATIC}} = 0.629$$

*Note:* In any of the previous examples of the conventional relationships for steady flow:

$$\left(\frac{T_{STAGNATION}}{T_{STATIC}}\right)^{\gamma/(\gamma-1)} = \frac{P_{STAGNATION}}{P_{STATIC}} = \left\{1 + \frac{\gamma-1}{2}M^2\right\}^{\gamma/(\gamma-1)}$$

can be invoked to relate static to stagnation conditions if this is required.

    The two auxiliary steps are yet to be described—one deals with scavenge processes, the other sets forth rules for computing port widths of single wave and scavenge processes.

### 7.7.4 Scavenging processes

A feature of some pressure-exchanger cycles is nominally waveless periods during which the fluid in the cell is changed for other fluid. Such processes are known as scavenging processes; an example of a scavenging process is the replacement of hot products of combustion by cool, fresh, air in gas generator type pressure exchangers.

    Figure 7.21 shows a typical scavenge process in which, in principle at least, there are no wave events during the period in which the cells move from $A$ to $B$.

    It has been demonstrated experimentally that the fluid within the cells slows down during scavenging. This has the effect of weakening the wave occurring at the termination of the process. This effect can be significant particularly when wide ports are employed and it is, therefore, necessary to make at least an approximate prediction of the decay of velocity during a scavenging process. While residual waves washing back and forth through the cells can influence the velocity distributions in the ports, viscous effects only are taken into account in the estimation of decay of velocity given here.

    Applying Newton's Second Law to the mass of fluid, $\delta m$, contained within an elementary length, $\delta x$, of a cell of any station between $A$ and $B$ of the scavenge

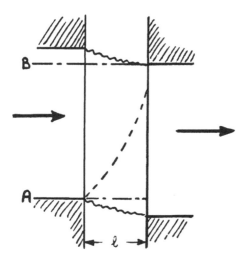

**Fig. 7–21** A low-pressure scavenging process showing the nominally waveless region $A \sim B$.

process shown in Fig. 7.21 and assuming that the static pressure along the axis of each cell is uniform (no waves):

$$\delta F_{\text{FRIC}} = -S(\delta m u) \tag{7.22}$$

where:

$S \equiv$ substantial derivative (Eq. 3.7, Chapter 3),
$\delta F_{\text{FRIC}} \equiv$ force due to friction, acting in a direction opposing flow, on the elemental mass $\delta m$.

But the fluid can be assumed to move through the cell as if it were incompressible (i.e., as if it were a solid rod slowing down with increasing time) hence for any element of mass $\delta m$ within the scavenge flow neither $\delta m$ or $u$ vary with displacement alone, hence:

$$\frac{\partial}{\partial x}(\delta m \, u) = 0 \tag{7.23}$$

and thus substituting the result (7.23) in (7.22) allows (7.22) to be rewritten:

$$\delta F_{\text{FRIC}} = -\frac{\partial}{\partial t}(\delta m \, u)$$

or since $\delta F_{\text{FRIC}}$ is independent of $x$;

$$\delta F_{\text{FRIC}} = -\frac{d}{dt}(\delta m \, u) \tag{7.24}$$

or, since an individual element of fluid will not change its mass as it travels along the length of a cell:

$$\delta F_{\text{FRIC}} = -\delta m \frac{du}{dt} \tag{7.25}$$

Expressing $\delta F_{\text{FRIC}}$ in terms of a friction factor $f$:

$$\delta F_{\text{FRIC}} = \left(\frac{\delta x}{d_h}\right) f \frac{\rho u^2}{2} A_{\text{CELL}} \tag{7.26}$$

where:

$\quad d_h \equiv$ hydraulic mean diameter of a cell,
$\quad A_{\text{CELL}} \equiv$ cross-sectional area of a cell

Thus from (7.25) and (7.26) and the substitution of $\delta m = \rho A_{\text{CELL}} \, \delta x$:

$$dt = -\frac{2d_h}{f} \frac{du}{u^2} \tag{7.27}$$

Equation (7.27) is independent of density and may, therefore, be integrated between $A$ and $B$. Thus taking $f$ as constant:

$$t_B - t_A = \frac{2d_h}{f} \left[ \frac{1}{u_B} - \frac{1}{u_A} \right] \tag{7.28}$$

Expressing (7.28) in terms of Mach numbers, substituting for $t$ from $t' \equiv a_{\text{REF}} t/l$ and rearranging:

$$M_B = \left(\frac{a_{\text{REF}}}{a_B}\right) \Big/ \left\{ \frac{a_{\text{REF}}}{a_A M_A} + \frac{\Delta t' l f}{2d_h} \right\} \tag{7.29}$$

where:

$\quad \Delta t' \equiv$ dimensionless time required for a cell to move from $A$ to $B$.

A representative value of $f$ for typical cell geometries has been found to be 0.025, this includes an allowance for flow entry loss.

Equation (7.29) is applicable whether it is assumed that the static pressure remains uniform throughout the scavenging process or, alternatively, the total pressure is constant with the static pressure gradually increasing to compensate for the effect of velocity decay during scavenge. For conditions typically prevailing in practice it matters but little which assumption is made.

### 7.7.5  Evaluation of port widths

The approximate (i.e., within limitations imposed by the assumptions listed in Section 7.7.1) widths of the ports can be established, in terms of $t'$, by means of the Approximate Method.

Application of continuity of mass to cells passing through the dotted boundaries of the flow zones for Cases 1 and 2 of Fig. 7.15 gives:

for Case 1:

$$m_2 + m_1 = m_4 + m_3 \tag{7.30}$$

for Case 2:

$$m_4 + m_1 = m_2 + m_3 \tag{7.31}$$

These equations were reformulated by expressing $m_2$ and $m_4$ in terms of the flow velocities in the cell ends and the length of time over which each cell communicates with a port, also $m_1$ and $m_3$ were expressed in terms of cell volume and the pressure and temperature within the cells. The following results were then finally obtained for the time, $\Delta t'_{\text{PORT(SW)}}$, for single wave processes as illustrated in Fig. 7.15. Thus:

$$\Delta t'_{\text{PORT(SW)}} = \frac{\pm\left\{1 - \dfrac{P_2}{P_4}\dfrac{T_4}{T_2}\right\}\dfrac{a_{\text{REF}}}{a_4}}{\left\{\dfrac{P_2}{P_4}\sqrt{\dfrac{T_4}{T_2}}\,M_2 - M_4\right\}} \tag{7.32}$$

or in alternative form:

$$\Delta t'_{\text{PORT(SW)}} = \frac{\pm\left\{1 - \dfrac{P_4}{P_2}\dfrac{T_2}{T_4}\right\}\dfrac{a_{\text{REF}}}{a_2}}{\left\{\dfrac{P_4}{P_2}\sqrt{\dfrac{T_2}{T_4}}\,M_4 - M_2\right\}} \tag{7.33}$$

The positive sign in Eqs. (7.32) and (7.33) applies to Case 1, the negative sign to Case 2. The equations are presented in the notation of Fig. 7.15.

The width, $\Delta t'$, of a scavenge port between stations $A$ and $B$ (see Fig. 7.21) can be obtained from the following relationship:

$$\frac{\Delta x}{l} = \frac{2}{f}\left(\frac{d_h}{l}\right)\log_e\frac{a_A M_A}{a_B M_B} \tag{7.34}$$

used in conjunction with Eq. (7.20); $\Delta x$ is the distance a particle moves along a cell during the interval $\Delta t'$ thus:

$$\Delta x = x_B - x_A \tag{7.35}$$

(Note: for a full scavenge $\Delta x = l$.)

It can, therefore, be seen that $\Delta x/l$ is the scavenge ratio when $x_A = 0$ (or 1.0 for a scavenge process in the negative $u$ direction).

Equation (7.34) was obtained by writing: $dx = u\,dt$, substituting for $dt$ in terms of $u$ and $du$ from Eqs. (7.25) and (7.26) and integrating the result between the limits $A$ and $B$.

It should be noted that for low pressure scavenging, $\Delta t'$ for scavenge is less than the total width of a port (see Figure 7.21) by an amount corresponding to

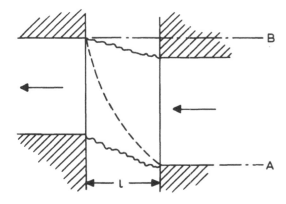

**Fig. 7–22** A high-pressure scavenging process showing the nominally waveless region $A \sim B$.

one single wave process. The opposite is true for a high pressure scavenging process as indicated in Fig. 7.22.

The finite width of cells has only a secondary influence on the total predicted port widths and relative phasings. Accordingly no special attention was paid to the effect of cell width on port widths and phasing. It should be remembered that the effect of cell width on performance is included in the basic cell filling and emptying data.

## 7.8  Practical Applications

Currently dynamic pressure-exchangers are available commercially, from one company, for supercharging applications. Tests have also been carried out by a number of workers with various experimental, prototype, units suitable for a number of other applications.

### 7.8.1  Superchargers

Dynamic pressure-exchanger superchargers available commercially are manufactured by Brown, Boveri and Company (now Asea Brown, Boveri) of Baden, Switzerland. These machines, which are identified by the name "Comprex" incorporate the results of an extensive development program. They are manufactured in sizes suitable for most automotive and earth-moving-machine applications to suit engines with outputs ranging from 42 to 450 kW.

Figure 7.23, which is reproduced through the courtesy of the Brown, Boveri Company, shows in a diagrammatic manner the application of a Comprex type pressure-wave supercharger (PWS) to a diesel engine represented as a symbolic single cylinder unit. Figure 7.24, also from Brown, Boveri, is a cut-away, or part-sectioned, diagram of a typical unit. The hot-gas high-pressure entry and the low-pressure outlet port are on the right. The low pressure air enters through the lower duct at the left-hand end and leaves, in the compressed state, through

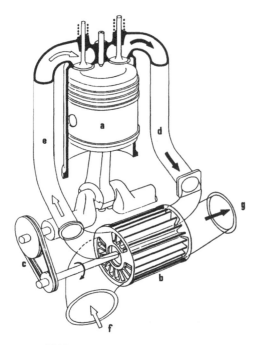

a   ENGINE
b   CELL WHEEL (ROTOR)
c   BELT DRIVE
d   HIGH PRESSURE GAS (G-HP)
e   HIGH PRESSURE AIR (A-HP)
f   LOW PRESSURE AIR (A-LP)
g   LOW PRESSURE GAS (G-LP)

**Fig. 7–23** Diagrammatic illustration of a Comprex pressure-wave supercharger (courtesy of Brown, Boveri and Co. Ltd.).

the upper air outlet duct. The rotor is normally driven by a belt, as is also implied in Fig. 7.23, from the engine crankshaft. An advanced version of the Comprex has been developed featuring an extruded, ceramic, self-driving, free-running rotor with mid-span cell-partition support (Zehner et al., 1989).

The stator and port arrangement of the Brown, Boveri machine differs somewhat from the simplistic configuration of Fig. 7.5. This is due to the need to cope with variable engine loads and speeds. To this end three pockets have been provided in the stator as shown in Fig. 7.25. It should be noted that in Fig. 7.25 the cells travel downward. The pockets serve to prevent flow reversals that would otherwise occur at low engine (and hence also low pressure-exchanger rotor) speeds and also under light load conditions at the design speed.

More specifically it has been shown by Croes (1979) that under high-load low-speed operating conditions the expansion pocket (ET in Fig. 7.25) serves

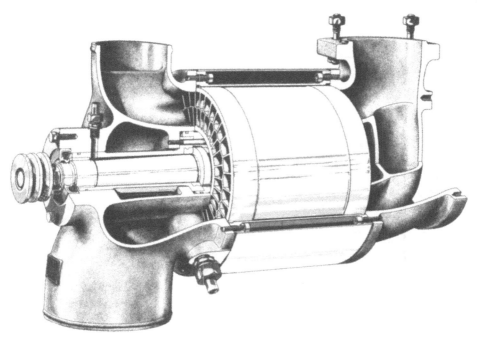

**Fig. 7–24** Cut-away diagram of a Comprex pressure-wave supercharger (courtesy of Brown, Boveri and Co. Ltd.).

to prevent flow reversal in the low-pressure tract while the compression pocket (KT in Fig. 7.25) inhibits flow reversals in the high-pressure tract. The gas pocket (GT in Fig. 7.25) serves to allow the maintenance of an adequate low-pressure scavenge flow under low load conditions at or close to the design speed.

A more thorough analysis of the functioning of each of the three pockets requires use to be made of the method-of-characteristics. Such an analysis was used by Croes (1979) in his explanation of the function of the pockets. It is also noteworthy that, in order to maximize the through-flow during low-pressure scavenge, the low-pressure outlet port closes *after* the low-pressure inlet port in contrast to the arrangement shown in Fig. 7.5.

The operational flexibility of the pressure-wave supercharger developed by Brown, Boveri is due, in large measure, to the special features described here. The benefit of these can be appreciated from Fig. 7.26 (Summerauer et al., 1978) that shows that the Comprex type machine is operationally much more responsive, under acceleration conditions, than a turbocharger.

An additional feature of interest is that the Brown, Boveri machines employ rotors featuring cells of unequal width. The purpose of this is to detune the characteristic noise normally made by pressure-exchangers. The unequal cell

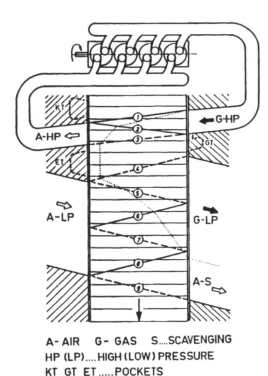

A- AIR   G- GAS   S....SCAVENGING
HP (LP)....HIGH (LOW) PRESSURE
KT GT ET.....POCKETS

**Fig. 7–25** Diagram of a Comprex supercharger on a distance ~ time plane showing the compression, expansion, and gas pockets. In this diagram the cells move in the downward direction (courtesy of Brown, Boveri and Co. Ltd.).

width features has contributed to not only making the machine less objectionable subjectively but also to reducing noise levels.

### 7.8.2   Gas generators

Pressure-exchangers of the gas generator type have been tested by Power Jets (R and D) Ltd. and by other companies.

Typical representative results obtained with a Power Jet gas generator are shown in Fig. 7.27. The unit was of the most simple type and did not, for example, feature transfer ports (Fig. 7.7).

Control was by varying the fuel flow quantity; there was no bleed-off of hot gas or compressed air from the HP circuit. The fall-off of the isentropic compression efficiency, $\eta_{COMP}$ (upper curve family), was due to hot gas mixing with the compressed air delivered into the HP loop this effect becoming ever more significant as the HP scavenge ratio (centre diagram) increased toward, or exceeded, unity. The estimate of $\eta_{COMP}$ was based on the temperature rise of the air (due to compression in the pressure-exchanger) by measuring stagnation temperatures at the LP inlet and HP outlet ports.

The lower curve family shows the bulk input (i.e., in effect the ratio of the

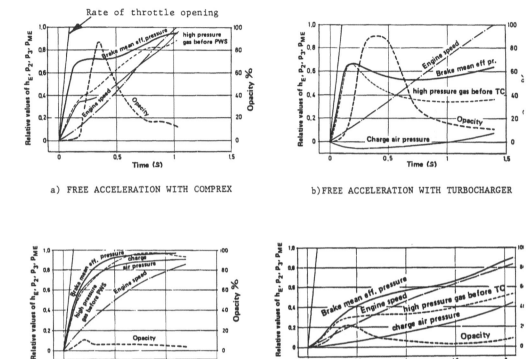

a) FREE ACCELERATION WITH COMPREX          b) FREE ACCELERATION WITH TURBOCHARGER

c) ENGINE ACCELERATION UNDER              d) ENGINE ACCELERATION UNDER
   LOAD WITH COMPREX                         LOAD WITH TURBOCHARGER

**Fig. 7–26** Transient performance of a Comprex supercharger compared with that of a conventional turbocharger (Summerauer et al., 1978).

HP circuit temperature rise due to combustion divided by the absolute temperature in the LP inlet port) versus pressure ratio. The straight dotted line is the theoretical, nondimensional, bulk input for a waveless (static) pressure-exchanger.

### 7.8.3   Coolers

At least two dynamic pressure-exchanger coolers were built by Power Jets (R and D) Ltd. These units were later operated as deep-mine environmental air-cooling devices. Figure 7.28 is an illustration of the first prototype unit. The pressure-exchanger and water-cooled heat-exchanger are in the foreground. The portion of the mine ventilating duct in which the unit was located is shown in the upper part of the photograph. The system employed the type of port arrangement illustrated in Fig. 7.6. The make-up airflow to the high-pressure circuit was tapped from the mine compressed air supply.

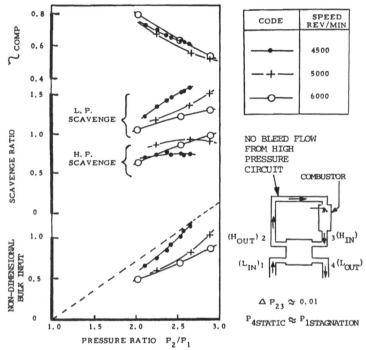

NOTE: ROTOR SPEED AT MEAN RADIUS OF CELLS = 8.5m/s per 1000rev./min.

**Fig. 7–27** Performance of a simple dynamic pressure-exchanger operating as a gas generator (with acknowledgments to the former Power Jets (R and D) Ltd.).

In each of the two prototype units the rotor was made self-driving by angling the high-pressure inlet to give the entering flow an incidence angle of approximately 10° to the cell walls at the operating speed. The coefficient of performance (cooling duty divided by the isentropic work of compression of the make-up air) was 1.2 for the second prototype unit. Noise levels were controlled by use of acoustic splitters that proved adequate to make the noise of the installed units insignificant compared with typical mine noise levels.

### 7.8.4  Equalizers

The performance of pressure-exchanger equalizers (Fig. 7.8) have been studied both theoretically and experimentally. Figures 7.29 to 7.33 inclusive present the results of theoretical predictions of equalizer performances for conditions ranging from idealized to those which closely approach the operation of real machines.

Considering the diagrams in sequence:

Fig. 7.29:  presents the idealized performance based on isentropic flow and the $u \sim a$ diagram concepts,

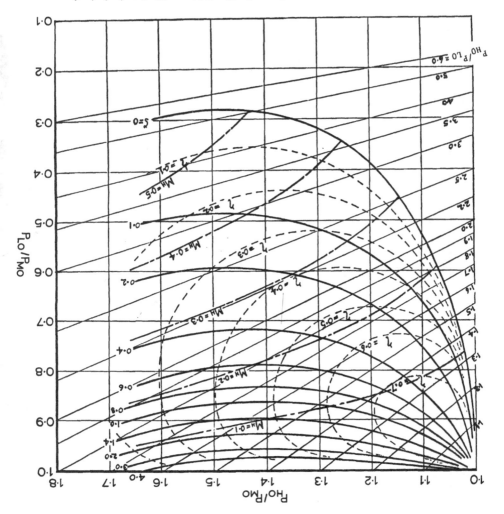

Fig. 7-33 The equalizer performance shown in Fig. 7.31 modified to include the influences of both leakage and friction.

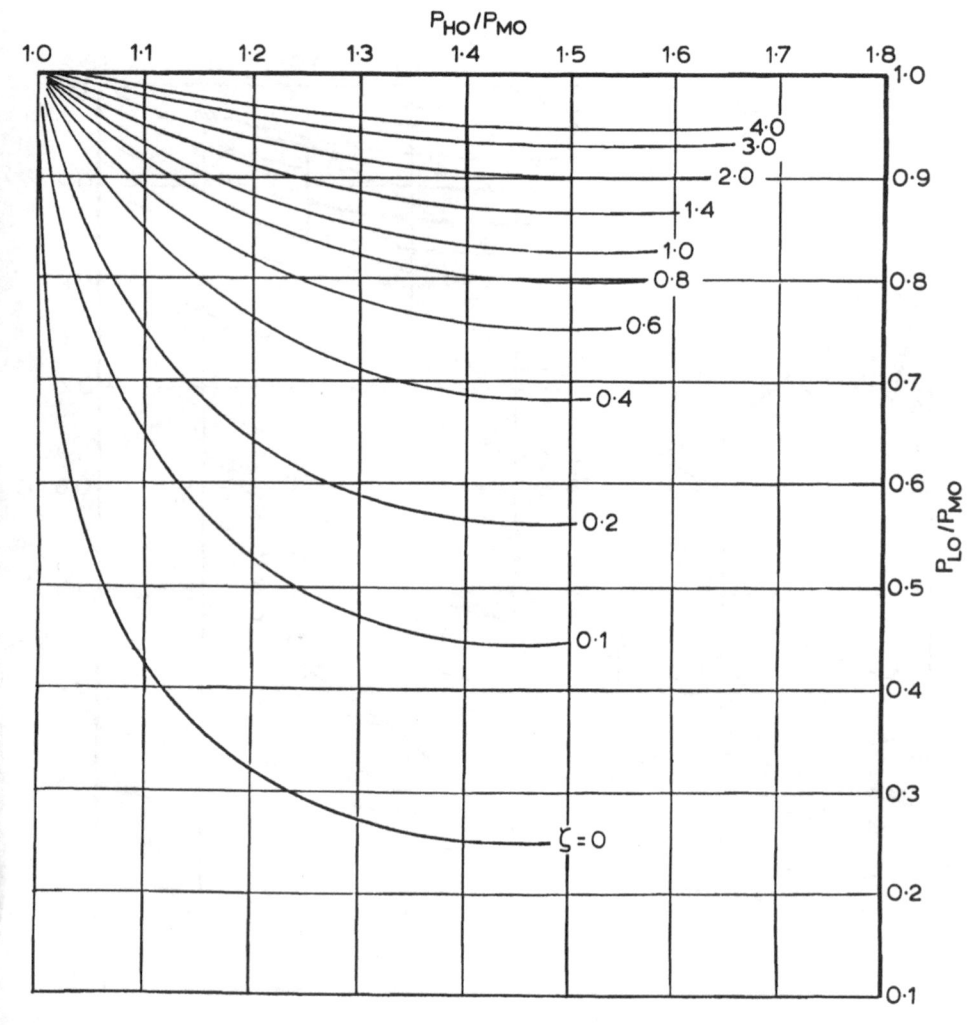

**Fig. 7–32** The equalizer performance in Fig. 7.3 modified to include the influence of leakage.

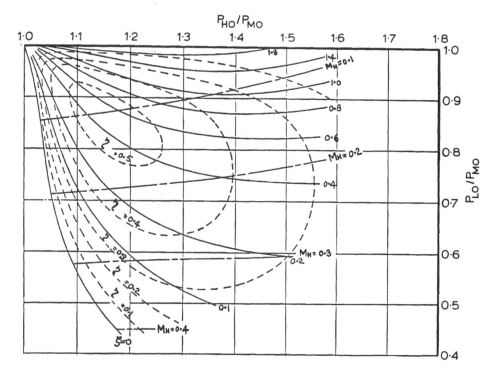

**Fig. 7–34** Performance of a pressure-exchanger equalizer with finite width cells
($\Delta t'_{CELL} \simeq 0.54$) predicted by means of the approximate procedure described in Section
7.7 ($\gamma = 1.4$).

Figure 7.35 shows an experimentally obtained performance map for an
equalizer in which $\Delta t'_{CELL} \approx 0.36$. A comparison of Figs. 7.33 and 7.35 shows a
fairly close agreement between the predicted and experimentally obtained
results. It should, however, be borne in mind that the experimental results were
obtained for a fixed geometry whereas the port geometry associated with Figs.
7.29 to 7.33 inclusive was adjusted to the theoretical optimum value at all points
on the performance plane.

The subscript "o" in the pressure terms in the pressure ratios on the axes
of the performance maps is merely to signify stagnation as distinct from static
conditions. The term $\zeta$ is defined thus:

$$\zeta \equiv \frac{C_{PL} \dot{m}_L T_{Lo}}{C_{PH} \dot{m}_H T_{Ho}}.$$

As can be seen from the performance curves, the equalizer is only suitable
for operation at low-pressure ratios. The combination, or product of the
isentropic efficiencies of compression and expansion, $\eta$, falls as the pressure
ratio increases.

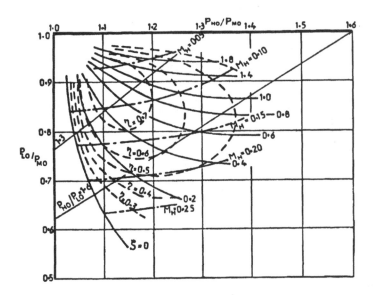

**Fig. 7–35** Experimentally obtained equalizer performance map, with $\Delta t'_{CELL} \simeq 0.36$, using air as the working fliuid ($\gamma = 1.4$).

More detailed studies of equalizers of the type described here are available (Kentfield, 1963; Kentfield, 1969). The performance of simplified equalizers with only one stator end-plate (see question 7.4) has been investigated both theoretically and experimentally by Ruf (1967).

### 7.8.5   Dividers

The performance characteristics of dividers have also been examined theoretically and experimentally. Much the same conclusions can be arrived at for the divider case regarding the relative importance of cell width, leakage and friction as for the equalizer. Furthermore, the agreement between theory and experiment is comparable for both the divider and equalizer. Accordingly, only an experimentally obtained performance curve (Fig. 7.36) has been included for a divider configuration of the type shown in Fig. 7.9. The experimental unit from which the test results were obtained is shown in Fig. 7.37.

As for the equalizer, the divider is, in the form tested, only suitable for operation at low pressure ratios. The parameter $\beta$ of Fig. 7.36 is defined thus:

$$\beta \equiv \dot{m}_H / \dot{m}_M.$$

Further experimentally obtained information is available (Kentfield, 1963; Kentfield, 1969) for dividers of the type depicted in Fig. 7.9. The use of machines of this kind as devices for reducing the pumping energy requirements of natural

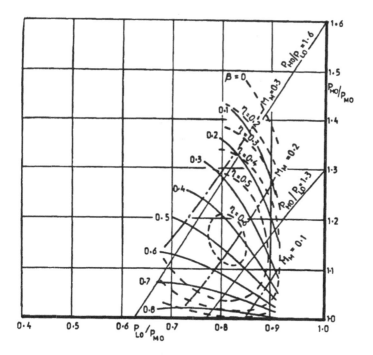

**Fig. 7–36** Experimentally obtained performance of a pressure-exchanger divider, with $\Delta t'_{CELL} \simeq 0.36$, using air as the working fluid ($\gamma = 1.4$).

gas pipelines has also been investigated (Kentfield and Barnes, 1976). This divider application, which was suggested by Barnes in the written discussion of a paper by the writer (Kentfield, 1969), is based on the use of a pressure divider as a substitute for a pressure-reducing valve, or governor, at a location in a natural gas pipeline where the line branches. Essentially, the flow passing to a customer connected to the low-pressure arm of the branch results in compression of the remainder of the gas flowing on in the high-pressure line.

The performance of simplified pressure-exchanger dividers featuring only a single stator end-plate has also been investigated both theoretically and experimentally by Ruf (1967). A divider with a single end-plate is illustrated diagrammatically in connection with question 7.4.

### 7.8.6   Wave superheater

A dynamic pressure-exchanger can also be configured to serve as a useful laboratory tool. An example of such an application was made by the Cornell Aeronautical Laboratory Incorporated of Buffalo, New York, now Arvin

**Fig. 7–37** The experimental pressure-exchanger divider test rig employed to obtain the results presented in Fig. 7.36.

Calspan. The Cornell unit was given the name "wave superheater" that reflects the function of the device.

In effect, the cells of the machine served as a series of shock tubes within each of which a charge was compressed, and subsequently discharged from the rotor, by a high-pressure driver gas flow of helium or hydrogen. The rotor was cooled, by a flow of coolant gas, during the low-pressure portion of the cycle, prior to being recharged with air.

The machine, which is illustrated in Figs. 7.38 and 7.39, is essentially a pressure-exchanger somewhat similar to the configuration shown in Fig. 7.5 with a very high applied pressure ratio of approximately 100. The output flow is the high-pressure, compressed, flow leaving the high-pressure outlet. Detailed

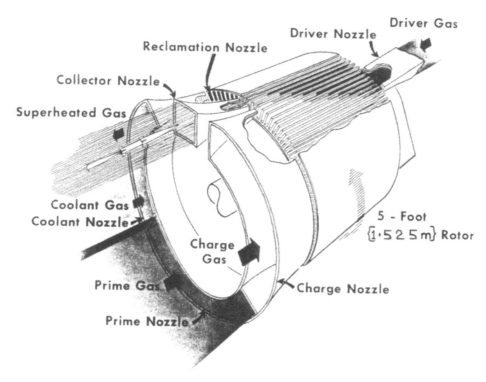

**Fig. 7–38** Diagrammatic illustration of the wave superheater pressure-exchanger (courtesy of Arvin Calspan).

descriptions of the wave superheater and its operating principles are available (Smith and Weatherston, 1958; Martin, 1960; Weatherston et al., 1959).

### 7.9 Optimum Cell Width and Rotor Speed

As indicated by Azoury (1966) the optimum cell width, $\Delta t'_{CELL}$, for dynamic pressure-exchangers measured in terms of dimensionless time is, when wall friction is ignored, approximately 0.5 for emptying processes and zero for filling events. Since pressure-exchanger cycles involve both filling and emptying processes this suggests a compromise optimum cell width of about $\Delta t'_{CELL} \simeq 0.25$. When allowance is made for cell wall friction, a more realistic value corresponds to $\Delta t'_{CELL} \simeq 0.35$.

Selection of a suitable value of the cell width in terms of dimensionless time is not sufficient to define the rotor design. The designer must also select a suitable rotor speed. Unlike a turbomachine the potential throughput of a pressure-exchanger is independent of rotor speed but is proportional to the port area provided. The selection, therefore, of a relatively low rotor speed will

**Fig. 7–39** Rotor of the wave superheater pressure-exchanger (courtesy of Arvin Calspan).

result in relatively narrow ports; this usually permits more than one such set of ports to be accommodated on a stator end plate. The selection of a relatively high rotor speed will, in the limit, permit the installation of only a single set of wide ports. To a first approximation the total port area will be the same for both situations. The choice of a relatively high rotor speed with, consequently, only a single set of ports simplifies the stator end-plate ducting but intensifies problems associated with the non-one-dimensionality of cell inlet flow, generated before the cell is fully open to the port, since the cell aspect ratio (i.e., the rotor length divided by the mean cell width) will be relatively small. A large cell aspect ratio, associated with a low rotor speed, will ensure that the entering flow will be restored to an essentially one-dimensional form before penetrating very far into the cell. Too large an aspect ratio will, however, introduce excessive losses due to cell-wall friction.

Clearly an optimum cell aspect ratio will minimize the losses due to the combined influences of flow entry and cell-wall friction. On the basis of somewhat incomplete evidence it would appear that an optimum cell aspect ratio is in the region of 20 suggesting that a rotor speed should be selected that ensures that this condition is met. The rotor length, and cell width, can be adjusted to ensure that the best possible use is made of the available stator end-plate area for accommodating ports.

An additional problem associated with the selection of a high rotor speed is the intensification, in proportion to the square of the rotor speed, of the tendency at an interface between gases of differing density, for the gas of high density to climb, due to centrifuging, on top of the gas of lower density. The cell-wall-partition mid-span support employed by Brown, Boveri (Zehnder et al., 1989) serves not only as a structural member but also to suppress mixing, at an interface, due to centrifuging. The selection of a low rotor speed is in conflict with the concept of configurations in which an attempt is made to combine, in one rotor, the functions of a pressure-exchanger and also a power turbine or turbocompressor.

## 7.10  Bibliography

A bibliography has been compiled of papers, patents, theses, etc., relating to pressure-exchangers. The bibliography, in which the entries are listed in chronological order, is included here to assist researchers, and others, and to illustrate the diversity of material available. The bibliography was compiled, in very large measure, by Thomas Lutz of the Swiss Federal Institute of Technology (E.T.H.) Zürich, Switzerland. The assistance of the Comprex Department of Asea Brown, Boveri is also gratefully acknowledged. Works referred to, specifically, in this chapter will be found listed in the Reference section at the end of the book: those that qualify appropriately can also be found in the Bibliography of Pressure-Exchanger Literature.

## Bibliography of Pressure-Exchanger Literature

| Year | Author, Authors, or Lead Author | Title and Publication |
|------|-------------------------------|----------------------|
| 1906 | Knauff, R. | British Patent No. 2819. |
| 1906 | Kanuff, R. | British Patent No. 8273. |
| 1913 | Burghard, H. | British Patent No. 19421. |
| 1928 | Lèbre, A. F. | British Patent No. 290 669. |
| 1928 | Burghard, H. | German Patent No. 485 386. |
| 1940 | Seippel, C. | Swiss Patent No. 225 426. |
| 1947 | Meyer, A. | "Recent Developments in Gas Turbines," *Mechanical Engineering*, Apr., pp. 273–277. |
| 1948 | Fauré | French Patent No. 935 362. |
| 1949 | Kantrowitz, A. et al. | Heat Engines Based on Wave Processes. Report of the Graduate School of Aeronautical Engineering, Cornell University, Ithaca, N.Y., Mar. 21. |
| 1950 | Barry, F. W. | "Introduction to the Comprex," *Journal of Applied Mechanics*, Mar., pp. 47–53 (originally presented as ASME paper 49-APM-18, 1949). |
| 1954 | Kantrowitz, A. | U.S. Patent No. 2,665,058. |
| 1955 | Pearson, R. D. and Hooton, G. E. | British Patent No. 843,911. |
| 1955 | Spalding, D. B. | Wave Effects in Pressure Exchangers, Part II. Power Jets (Research and Development) Ltd., Rept. No. 2202/x 41. |
| 1955 | Poggi, L. | "The Theory of Semi-Static Pressure Exchangers," *Selected Papers on Engineering Mechanics*, von Kàrmàn Tribute Vol. 83, Butterworth, London. |
| 1956 | Spalding, D. B. | Filling, Emptying, and Transfer Processes in a Pressure Exchanger of Finite Cell Width. Power Jets (Research and Development) Ltd., Rept. No. 2222/x 46. |
| 1956 | Spalding, D. B. | Theoretical Considerations on Leakage Effects in Pressure Exchangers. Power Jets (Research and Development) Ltd., Rept. No. 2224/x 47. |
| 1956 | Jendrassik, G. | U.S. Patent No. 2,757,509. |
| 1958 | Burri, H. U. | "Nonsteady Aerodynamics of the Comprex Supercharger," ASME Paper 58-GTP-15. |
| 1958 | Spalding, D. B. | British Patent No. 799,143. |
| 1958 | Jendrassik, G. | German Patent No. 1,030,506. |

| 1958 | Spalding, D. B. | A Note on Pressure Equalizers and Dividers. Power Jets (Research and Development) Ltd., Rept. No. 2251/Px 3. |
|---|---|---|
| 1958 | Smith, W. E. and Weatherston, R. C. | Studies of Prototype Wave Superheater Facility for Hypersonic Research. Cornell Aero. Lab. Rept. No. HF-1056-A-1, AFSOR TR 58-158. |
| 1958 | Barnes, J. A. and Spalding, D. B. | "The Pressure Exchanger. Its Operational Principles and Potentialities," *The Oil Engine and Gas Turbine*, Feb., p. 364. |
| 1958 | Berchtold, M. and Gardiner, F. J. | "The Comprex, A New Concept of Diesel Supercharging," Presented at the Gas Turbine Power Conference and Exhibit, Washington, Mar. 2–6. |
| 1958 | Berchtold, M. | "The Comprex Diesel-Supercharger," SAE Paper No. 63A. |
| 1959 | Berchtold, M. and Gull, H. P. | "Road Performance of a Comprex-Super-charged Diesel Truck," SAE Paper No. 118U. |
| 1959 | Weatherston, R. C. et al. | Gas Dynamics of a Wave Superheater Facility for Hypersonic Research and Development. Cornell Aero. Lab. Rept. No. AD-1118-A-1, AFSOR TN 59-107. |
| 1959 | Berchtold, M. | U.S. Patent No. 2,867,981. |
| 1960 | Azoury, P. H. | "The Dynamic Pressure Exchanger—Gas Flow in a Model Cell," Ph.D. thesis, University of London. |
| 1960 | Martin, J. F. | "New Tool for Research," *Research Trends*, Vol. 7, No. 4, Cornell Aeronautical Laboratory Inc., New York. |
| 1961 | Azoury, P. H. | Temperature Discontinuities in a Pressure Exchanger of Finite Cell Width. Power Jets (Research and Development) Ltd., Rept. No. 2262/Px 9. |
| 1961 | Azoury, P. H. | High-Pressure and Low-Pressure Scavenge in a Pressure Exchanger of Finite Cell Width. Power Jets (Research and Development) Ltd., Rept. No. 2263/Px 10. |
| 1963 | Kentfield, J. A. C. | "An Examination of the Performance of Pressure Exchanger Equalisers and Dividers," Ph.D. thesis, University of London. |
| 1963 | Carpenter, J. E. | "The Wave Superheater Hypersonic Tunnel," *Research Trends*, Cornell Aeronautical Laboratory Inc., New York. |
| 1964 | —— | "Pressure Exchanger Progress," *The Oil Engine and Gas Turbine*, Vol. 32, No. 371, pp. 35. |

1966    Azoury, P. H.    "An Introduction to the Dynamic Pressure Exchanger," *Proc. Inst. Mech. Engrs.*, Vol. 180, Pt. 1, No. 18.

1966    Pfenninger, H.    "The Evolution of the Brown Boveri Gas Turbine," Gas Turbine Conference, Zurich, Mar. 13–17 (BBC-Sunderdruck 3217 E).

1967    Weatherston, R. C. and Hertzberg, A.    "The Energy Exchanger; A New Concept for High Efficiency Gas Turbine Cycles," ASME *Journal of Engineering for Power*, Vol. 89.

1967    Lutz, T. W. and Scholz, R    "Ueber die Aufladung von Fahrzeug-Dieselmotoren mittels des Comprex-Druckaustauschers," *Motortech. Z.*, Vol. 28, No. 5, p. 174.

1967    Suchow, E. J.    "Experimentelle Untersuchung der Arbeitsweise des Druckaustauschers," *Energomashinostroenie*, Vol. 13, No. 8, p. 23.

1967    Ruf, W.    "Berechnung und Versuche au Druckwellenmaschinen unter besonderer Berücksichtigung des Druckteilers und Injektors," Dissertation Nr. 3937, Eidgen. Techn. Hochschule, Zürich.

1968    Lutz, T. W. and Scholz, R.    "Supercharging Vehicle Engines by the Comprex System," The Inst. Mech. Engrs.— Paper presented Oct. 16.

1968    Wunsch, A.    "Zum Stand der Entwicklung von gasdynamischen Druckwellenmaschinen für die Aufladung von Dieselmotoren," *Brown Boveri Mitteilungen*, Vol. 55, No. 8, p. 440.

1968    Kentfield, J. A. C.    "An Approximate Method for Predicting the Performance of Pressure Exchangers," ASME Paper 68-WA/FE-37.

1969    Hörler, H. U.    "Abschätzung der Verluste in instationär gasdynamischen Kanaltrommel-Drucktauscher," Dissertation Nr. 4402, Eidgen. Techn. Hochscule, Zürich.

1969    Kentfield, J. A. C.    "The Performance of Pressure-Exchanger Dividers and Equalizers," ASME *Journal of Basic Engineering*, Vol. 91, Series D, No. 3, pp. 361–368.

1969    Köbberling, R.    Comprex-Konkurrenz für den Turbolader Lastauto-Omnibus, Vol. 46, No. 12, p. 35.

1970    Wunsch, A.    "Aufladung von Fahrzeugdieselmotoren mit dem Abgasturbolader und mit der Druck-wellenmaschine Comprex," *Motortech. Z.*, Vol. 31, No. 1.

| 1970 | Schneikart, H. | "Comprex: Atemhilfe für den Dieselmotor," *Omnibus Revenue*, Vol. 21, No. 2. |
|------|---------------|------|
| 1971 | Zehnder, D. | "Calculating Gas Flow in Pressure-Wave Machines," *Brown Boveri Review*, No. 4/5/71, 4019 E. (6.71-500). |
| 1971 | Wunsch, A. | "Fourier-Analysis Used in the Investigation of Noise Generated by Pressure Wave Machines," *BBC-Revenue*, 4/5/71. |
| 1971 | Zubatov, N. | "Direct Energy Exchange and its Application to MHD-Installations," *Thermotechnical Problems of Direct Energy Conversion*, Scientific Thought Press, Kiev, USSR, Issue 2, pp.37–44. |
| 1971 | Zubatov, N. | "Thermodynamic Analysis of Some MHD-Installation Cycles with Energy Exchangers," *Thermotechnical Problems of Direct Energy Conversion*, Scientific Thought Press, Kiev, USSR, Issue 2, pp. 44–51. |
| 1972 | Zubatov, N. | "Energy Exchanger Equilibrium Wall Temperature at MHD-Installation Cycles," *Thermotechnical Problems of Direct Energy Conversion*, Scientific Thought Press, Kiev, USSR, Issue 3, pp. 56–64. |
| 1972 | Zubatov, N. | "Energy Exchanger's Frame and Efficiency as Function of MHD-Installation Power Output," *Thermotechnical Problems of Direct Energy Conversion*, Scientific Thought Press, Kiev, USSR, Issue 3, pp. 64–71. |
| 1973 | Jenny, E. and Bulaty, T. | "Die Druckwellenmaschine Comprex als Oberstufe einer Gasturbine," Part 1 in *MTZ*, Vol. 34, No. 10, pp. 329–335 and Part 2 in *MTZ*, Vol. 34, No. 12, pp. 421–425. |
| 1973 | Zubatov, N. | "Energy Exchanger's Static Characteristics at Open-Cycle of MHD-Installations," *Thermotechnical Problems of Direct Energy Conversion*, Scientific Thought Press, Kiev, USSR, Issue 4, pp. 97–104. |
| 1974 | Haefeli, R. | "Der Comprex, eine Möglichkeit zur Aufladung von Dieselmotoren," *Formel D*, No. 35, Jan. |
| 1974 | Berchtold, M. and Lutz, T. W. | "A New Small Power-Output Gas Turbine Concept," ASME Paper 74-GT-111. |
| 1974 | Berchtold, M. | "The Comprex Pressure Exchanger. A New Device for Thermodynamic Cycles," JSAE Paper, Tokyo, Japan, Aug. |

| 1974 | Rixmann | "MAN-Lastwagen-Dieselmotoren mit Comprex-Aufladung," *MTZ*, Vol. 35, No. 8. |
|------|---------|-------------|
| 1974 | Mayer, A. | "Le Procédé Comprex Nouveau Système de Suralimentation pour Moteur Diesel de Véhicules," *Ing. de l'automobile*, Oct. |
| 1975 | Eisele, E. et al. | "Experience with Comprex Pressure Wave Supercharger on a High Speed Passenger Car Diesel Engine," SAE Paper No. 750 334. |
| 1975 | Dörfler, P. | "Comprex Supercharging of Vehicle Diesel Engines," SAE Paper No. 750 335. |
| 1975 | Eisele, E. et al. | "Investigations into the Use of a Comprex-Supercharger System on a High-Speed Diesel Car Engine," *MTZ*, Vol. 36, No. 3. |
| 1976 | Coleman, R. C. and Weber, H. E. | U.S. Patent No. 3,811,796. |
| 1976 | Kentfield, J. A. C. and Barnes, J. A. | "The Pressure Divider: A Device for Reducing Gas-Pipe-Line Pumping-Energy Requirements," *Proceedings*, 11th Intersociety Energy Conversion Engineering Conference, AIChE, pp. 636–643 |
| 1977 | Croes, N. | "The Principle of the Pressure-Wave Machine as Used for Charging Diesel Engines," *Proceedings*, 11th International Symposium on Shock Tubes and Waves, Univ. of Washington Press, pp. 36–55. |
| 1977 | Kirchhofer, H. | "Aufladung von Fahrzeugdieselmotoren mit Comprex," *Automobil-Industries*, 1/77 DK 621.436.629.1. |
| 1977 | Kamo, R. | "Performance and Sociability of Comprex Supercharged Diesel Engine," ASME Paper 77-DGP-4. |
| 1978 | Schwarzbauer, G. E. | "Turbocharging of Tractor Engines with Exhaust Gas Turbochargers and the BBC-Comprex," I.Mech.E. Paper C69/78. |
| 1978 | Summerauer, I. et al. | "A Comparative Study of the Acceleration Performance of a Truck Diesel Engine With Exhaust-Gas Turbocharger and With Pressure Wave Supercharger Comprex," I.Mech.E. Paper C70/78. |
| 1979 | Croes, N. | "Die Wirkungsweise der Taschen des Druck-wellenladers Comprex," *MTZ*, Vol. 40 DK 621.43.052. |

| 1979 | Zumdieck, J. F. et al. | "The Fluid Dynamic Aspects of an Efficiency Point Design Energy Exchanger," Presented at the 12th International Symposium on Shock Tubes and Waves, Jerusalem, July. |
|------|------------------------|-----------------------------------------------------------------------------------------------------------------------------------------------------------------------------------------|
| 1980 | Rose, P. H. | "Potential Applications of Wave Machinery to Energy and Chemical Processes," *Proceedings*, 12th International Symposium on Shock Tubes and Waves, Magnes Press. |
| 1980 | Thayer, W. J. et al. | "Measurements and Modelling of Energy Exchanger Flow," *Proceedings*, 15th Intersociety Energy Conversion Engineering Conference, Seattle, Washington, Aug., p. 2368. |
| 1981 | Berchtold, M. | "Two Stage Supercharging with Comprex," CIMAC, Helsinki D111. |
| 1981 | Korhonen, V. | "Pressure Wave Supercharging Improves Engine Characteristics," *Diesel & Gas Turbine*, CH-T 123190E. |
| 1981 | Mayer, A. | "Comprex Supercharger for Vehicle Diesel Engine," *Diesel & Gas Turbine*, CH-T 123200E. |
| 1981 | Thayer, W. J. et al. | Energy Exchanger Performance and Power Cycle Evaluation: Experiments and Analysis. Final Report DOE/ER/01084-1, April. |
| 1981 | Thayer, W. J. and Zumdieck, J. F. | "A Comparison of Measured and Computed Energy Exchanger Performance," Presented at the 13th Symposium on Shock Tubes and Waves, Buffalo, New York, July 6–9. |
| 1982 | Wallace, F. J. | "Comprex Supercharging Versus Turbo-charging of a Large Truck Diesel Engine," I.Mech.E. Paper 82/C39. |
| 1982 | Jenny, E. | "Pressure Wave Supercharging of Passenger Car Diesel," I.Mech.E. Paper 4/82/C44. |
| 1982 | Mayer, A. and Schruf, G. M. | "Practical Experience with the Pressure Wave Supercharger Comprex on Passenger Cars," I.Mech.E. Paper C110/82. |
| 1982 | Shreeve, R. P. et al. | Wave Rotor Technology Status and Research Report. NPS 67-82-014PR, Naval Postgraduate School, Monterey, Nov. |
| 1983 | Gyarmathy, G. | "How Does the Comprex Pressure-Wave Supercharger Work?" SAE Paper 830234. |
| 1983 | Pearson, R. D. | "A Pressure Exchange Engine for Burning 'Pyroil' as the End User in a Cheap-power from Biomass System," CIMAC Conference Paper, Paris. |
| 1984 | Zehnder, G. and Mayer, A. | "Comprex Pressure Wave Supercharging for Automotive Diesels," SAE Paper 840 132. |

| 1984 | Coleman, R. R. | Wave Engine Technology Development. Final Report prepared by General Power Corporation for AFWAL (Contract No. AFWAL-TR-83-2095), Jan. |
|------|----------------|------|
| 1184 | Taussig, R. T. | "Wave-Rotor Turbofan Engines for Aircraft," *Mechanical Engineering*, ASME, Novber. |
| 1985 | Kantrowitz, A. | "Wave Engines," *Proceedings*, ONR/NAVAIR Wave Rotor Research and Technology Workshop, Naval Postgraduate School, Monterey, pp. 5–8. |
| 1985 | Kentfield, J. A. C. | "The Pressure Exchanger; An Introduction Including a Review of the Work of Power Jets (R & D) Ltd.," *Proceedings*, ONR/NAVAIR Wave Rotor Research and Technology Workshop, Naval Postgraduate School, Monterey, pp. 9–49. |
| 1985 | Berchtold, M. | "The Comprex," *Proceedings*, ONR/NAVAIR Wave Rotor Research and Technology Workshop, Naval Postgraduate School, Monterey, pp. 50–74. |
| 1985 | Matthews, L. | "The Gas Dynamics of Pressure Wave Superchargers," *Proceedings*, ONR/NAVAIR Wave Rotor Research and Technology Workshop, Naval Postgraduate School, Monterey, pp, 75–85. |
| 1985 | Thayer, W. J. | "The MSNW Energy Exchanger Research Program," *Proceedings*, ONR/NAVAIR Wave Rotor Research and Technology Workshop, Naval Postgraduate School, Monterey, pp. 86–115. |
| 1985 | Moritz, R. | "Rolls-Royce Study of Wave Rotors (1965–70)," *Proceedings*, ONR/NAVAIR Wave Rotor Research and Technology Workshop, Naval Postgraduate School, Monterey, pp. 116–124. |
| 1985 | Pearson, R. | "A Gas Wave-Turbine Engine Which Developed 35 hp and Performed Over a 6:1 Speed Ratio," *Proceedings*, ONR/NAVAIR Wave Rotor Research and Technology Workshop, Naval Postgraduate School, Monterey, pp. 125–170. |
| 1985 | Mathur, A. | "A Brief Review of the G.E. Wave Engine Program 1958–63," *Proceedings*, ONR/NAVAIR Wave Rotor Research and Technology Workshop, Naval Postgraduate School, Monterey, pp. 171–193. |

1985    Weber, H. E.              "Shock-Expansion Wave Engines—New
                                 Directions for Power Production," *Proceedings*,
                                 ONR/NAVAIR Wave Rotor Research and
                                 Technology Workshop, Naval Postgraduate
                                 School, Monterey, pp. 194–214.

1985    Mathur, A.               "Design and Experimental Verification of
                                 Wave Rotor Cycles," *Proceedings*,
                                 ONR/NAVAIR Wave Rotor Research and
                                 Technology Workshop, Naval Postgraduate
                                 School, Monterey, pp. 215–228.

1985    Eidelman, S.             "Gradual Opening of Axial and Skewed
                                 Passages in Wave Rotors," *Proceedings*,
                                 ONR/NAVAIR Wave Rotor Research and
                                 Technology Workshop, Naval Postgraduate
                                 School, Monterey, pp. 229–249.

1985    Wilson, D. G.            "Wave Rotors as Substitutes for Gas-Turbine
                                 Diffuser-Combustor Systems," *Proceedings*,
                                 ONR/NAVAIR Wave Rotor Research and
                                 Technology Workshop, Naval Postgraduate
                                 School, Monterey, pp. 250–260.

1985    Zubatov, N.              "Application of Wave Energy Exchanger to
                                 Industrial and Energy Systems," *Proceedings*,
                                 ONR/NAVAIR Wave Rotor Research and
                                 Technology Workshop, Naval Postgraduate
                                 School, Monterey, pp. 261–270.

1985    Rostafinski, W.          "Comparison of the Wave-Rotor-Augmented
                                 to the Detonation-Wave-Augmented Gas
                                 Turbine," *Proceedings*, ONR/NAVAIR Wave
                                 Rotor Research and Technology Workshop,
                                 Naval Postgraduate School, Monterey,
                                 pp. 271–283.

1985    Berchtold, M.            "The Comprex as a Topping Spool in a Gas
                                 Turbine Engine for Cruise Missiles,"
                                 *Proceedings*, ONR/NAVAIR Wave Rotor
                                 Research and Technology Workshop, Naval
                                 Postgraduate School, Monterey, pp. 284–290.

1985    Taussig, R.              "Wave Rotor Turbofan Engines for Aircraft,"
                                 *Proceedings*, ONR/NAVAIR Wave Rotor
                                 Research and Technology Workshop, Naval
                                 Postgraduate School, Monterey, pp. 291–328.

1985    Pearson, R.              "Performance Predictions for Gas Turbine-
                                 Wave Engines Including Practical Cycles with
                                 Wide Speed Range," *Proceedings*,
                                 ONR/NAVAIR Wave Rotor Research and
                                 Technology Workshop, Naval Postgraduate
                                 School, Monterey, pp. 329–378.

| 1985 | Pearson, R. | "Pressure Exchangers and Pressure Exchange Engines," Chapter 16 of *Internal Combustion Engines*, Vol. 2, Benson Memorial Volumes (Editors: Horlock, J. H. and Winterbourne, D. E.), Oxford University Press. |
|---|---|---|
| 1986 | Spinnler, F. and Jaussi, F. A. | "The Fully Self-Regulated Pressure Wave Supercharger Comprex for Passenger Car Diesel Engines," I.Mech.E. Paper C124/86. |
| 1987 | Jenny, E. et al. | "The Comprex Pressure Wave Supercharger," *Brown Boveri Review*, Aug. |
| 1988 | Hiereth, H. and Wiltham, G. | "New Results of Passenger Car Diesel Engine Pressure Wave Supercharged With and Without a Particulate Trap," SAE Paper 880005. |
| 1988 | Mayer, A. and Pauli, E. | "Emissions Concept for Vehicle Diesel Engines Supercharged with Comprex," SAE Paper 880008. |
| 1988 | Mayer, A. | "The Free Running Comprex—A New Concept for Pressure Wave Supercharger," SAE Document PC 55. |
| 1988 | Tatsuttomi, Y. et al. | "The Design and Development of the 626 RF PWS Diesel Engine," 3rd Supercharging Conference, Zurich, Sept. |
| 1989 | Mayer, A. et al. | "Extruded Ceramic – A New Technology for the Comprex-Rotor," SAE Paper 890425. |
| 1989 | Zehnder, G. et al. | "The Free Running Comprex," SAE Paper 890 452 (also SP-780). |
| 1989 | Hiereth, H. | "Car Tests With a Free-Running Pressure-Wave Charger—A Study for an Advanced Supercharging System," SAE Paper 890453 (also SP-780). |
| 1989 | Hitomi, M. et al. | "The Characteristics of Pressure Wave Super-charged Small Diesel Engine," SAE Paper 890 454 (also SP-780). |

## Problems

7.1  Show that, for the ideal static (Lèbre) pressure-exchanger gas-generator illustrated in Fig. 7.40, when there is no bleed from the HP loop the dimensionless heat input $q_{IN}$ required to maintain a prescribed value of $P_H/P_L$ approaches zero as the number, $n$, of transfer passages approaches infinity. Show also that $q_{IN} = \{P_H/P_L - 1]$ when $n = 0$.

Assume that the modulus of the change of pressure $|\delta P|$, within the cells as they pass a port is the same for all ports with the exception of the LP and HP scavenge ports in which there is zero change of pressure.

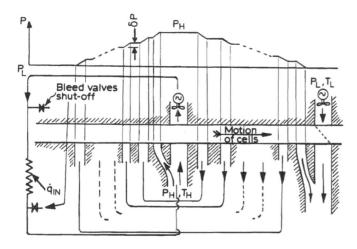

**Fig. 7–40** Diagram for question 7.1.

Pressure losses in the HP loop, LP scavenge zone, and transfer passages are to be ignored.

*Note:*

$$q_{IN} \equiv \frac{\text{(heat supplied to HP loop heater)/cell}}{\text{internal energy of contents of one cell after LP scavenge}}$$

7.2  Sketch the $u' \sim a'$ diagrams for the following cycles operating isentropically at idling conditions:

   a. Gas generator (Fig. 7.5),

   b. Gas generator with two transfer passages (Fig. 7.7).

   Compare $P_H/P_L$ for (a) with that for (b) when $\gamma = 1.4$, $a'_L = 1.0$, $u'_H = 0.4$, $u'_L = 0.1$ and, for (b) only, $u'_{TRANSFER PORTS} = 0.3$.

7.3  Sketch typical $u' \sim a'$ diagrams for the variants of the pressure divider and pressure equalizer configurations shown in Fig. 7.41. No attempt has been made to distinguish between compression and expansion waves in Fig. 7.41.

7.4  Sketch typical $u' \sim a'$ diagrams for the simplified versions of the divider and equalizer shown in Fig. 7.42. A special feature of each of these devices is that only one stator end-plate is employed.

7.5  It is proposed to simulate the combustion-in-cell device (Fig. 7.10) with a "cold rig" in which there is no combustion. The equivalent "cold rig" is shown in Fig. 7.43. The spark plug has, therefore, been replaced by a filling port. Estimate, on an isentropic basis, for the case in which $M_{M OUT} = 0.5$, $M_{L IN} = 0.3$ the:

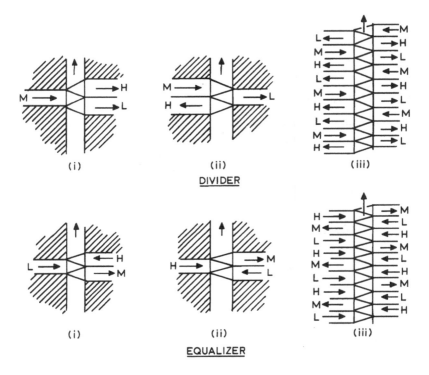

Fig. 7–41 Diagram for question 7.3.

a. stagnation pressure ratio $P_{oM\,OUT}/P_{oL\,IN}$,

b. stagnation pressure of the fluid supplied to the filling port as a fraction of of that entering the $L_{IN}$ port.

Assume that $\gamma = 1.4$.

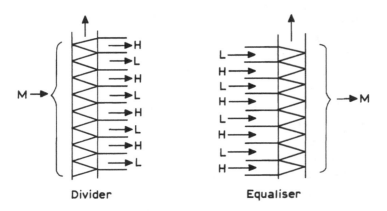

Fig. 7–42 Diagram for question 7.4.

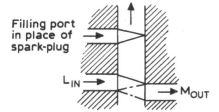

Fig. 7–43 Diagram for question 7.5.

7.6  Three proposals have been made for pressure-exchangers suitable for use as simple, cheap, single-rotor jet propulsion units for light aircraft. Figure 7.44 shows the proposals on the $x \sim t$ plane. Draw $u' \sim a'$ diagrams illustrating the isentropic operation of corresponding "cold" cycles in

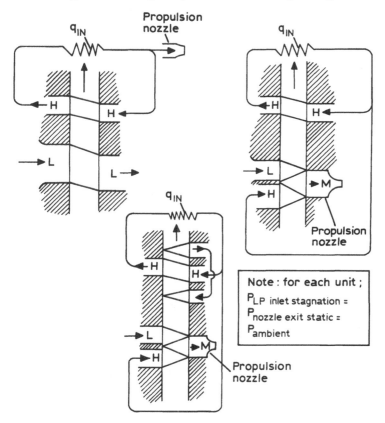

Fig. 7–44 Diagram for question 7.6.

which heat addition is replaced by mass addition in the HP circuits, thereby simulating the operation of the depicted units.

7.7  Present on a plane the ordinate of which represents the stagnation pressure ratio $P_{oH}/P_{oM}$, and the abscissa, the stagnation pressure ratio $P_{oL}/P_{oM}$, the performance characteristic of an idealized isentropic divider (Fig. 7.9) for which $\beta = 0.5$ [where $\beta \equiv \dot{m}_H/\dot{m}_M$]. Also locate the $\beta = 0$ and $\beta = 1.0$ curves and the intersection of the $M_M = 1.0$ contour with both of these: $\gamma = 1.4$.

7.8  Present, on a stagnation pressure ratio $P_{oL}/P_{oM}$ versus $P_{oH}/P_{oM}$ plane, the performance characteristic of an idealized isentropic equalizer (Fig. 7.8) for which $\zeta = 1.0$ [where $\zeta = \dot{m}_L T_{oL}/\dot{m}_H T_{oH}$]. Also draw on the diagram the $\zeta = 0$ and $\zeta \to \infty$ characteristics and mark the intersection of the $M_M = 1.0$ contour with both of these: $\gamma = 1.4$.

7.9

  i. Suggest leakage models for the configurations illustrated in Fig. 7.45.

  ii. Show how the leakage model for a filling process, Fig. 7.13, can be improved to take into account an additional source of leakage, present in the configuration shown in Fig. 7.13, without breaking any of the rules re leakage paths crossing the boundary of the wave diagrams.

  iii. Derive, from the analysis given in Section 7.6.3, the following expression relation to the effect of simple leakage:

$$\left[1 - \frac{P_c^*}{P_c}\right] = \gamma \Delta t'_{CELL} \, \Sigma \, \frac{A_l P_l}{A_{CELL} P_c} \, a'_l C_S \Phi_{l,x}$$

7.10

  i. Show, for a scavenge process, that with a loss of pressure external to the cells the equivalent of $n$ (local) dynamic heads in the cells expressions (7.29) and (7.34) of section (7.7) become:

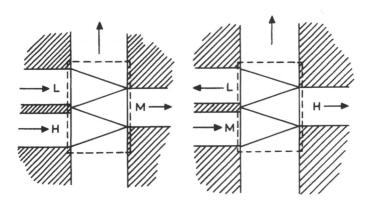

**Fig. 7–45** Diagram for question 7.9.

$$M_B = \left(\frac{a_{\text{REF}}}{a_B}\right) \bigg/ \left\{\frac{a_{\text{REF}}}{a_A M_A} + \frac{\Delta t'}{2}\left[\frac{fl}{d_h} + n\right]\right\}$$

and

$$\frac{\Delta x}{l} = \frac{2}{\left[\dfrac{fl}{d_h} + n\right]} \log_e\left[\frac{a_A M_A}{a_B M_B}\right]$$

ii. Evaluate the approximate geometry of a mine cooler, by means of the Approximate Method, in which the static pressures in the HP inlet and outlet ports are each 1.8 times the (uniform) static pressure in the LP inlet. The static temperature of the fluid entering the HP inlet is equal to the (uniform) static temperature of that entering the LP inlet port. The HP scavenge ratio is unity; the LP scavenge ratio is 1.2. Assume that no diffuser is fitted to the LP outlet and the consequent loss of pressure in the LP scavenge tract is one dynamic head (i.e., $n = 1$ in the formulas of the equations in part (i) of this equation). Establish the pressure is the cells between the end of LP and the beginning of HP scavenge and at the beginning of LP scavenge.

Assume that $\Delta t'_{\text{CELL}} \simeq 0.54$ for all ports and that the working fluid is (dry) air $\gamma = 1.4$.

Note: A point of interest is that the answer to this question yields a configuration very similar to that of an actual mine cooler, operating at similar pressure ratios, actually built and operated in a gold mine in the Republic of South Africa.

7.11 A mine cooler pressure-exchanger, as shown in Fig. 7.46, is to be designed. A requirement is that the static pressures in the HP inlet and outlet ports should each be 2.5 atmospheres. The static pressure in the LP inlet port is 1.0 atmosphere.

The fluid in the HP loop is cooled down to the static temperature of the ingoing LP air. The stagnation temperature of the make-up air entering the HP loop is equal to that of the HP flow leaving the cooler. The HP and LP scavenge ratios are (each) unity. Use the approximate procedure to make a preliminary design of the pressure-exchanger. Evaluate the width and phase of all four ports. Establish the Mach numbers at the leading and trailing edges of each port also the pressures within the cells before and after LP scavenge (or after and before HP scavenge). Ignore leakage effects related to lands between ports not taken into account implicitly by the prediction procedure. Recommend practical land widths in order to maximize the throughput of the machine. The friction factor, $f$, $= 0.024$; $l/d_h = 10$.

Suggested sequence:

i. Assume a pressure at $a$ of Fig. 7.46.

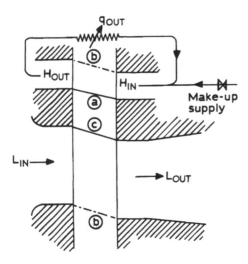

**Fig. 7–46** Diagram for question 7.11.

ii. Evaluate HP scavenging on the basis of the assumed value of $P_a$ and the specified data,

iii. On the basis of the pressure at $b$ derived from step (ii) proceed to evaluate LP scavenge taking the specified data into account,

iv. Does the final cell pressure $c$ derived from LP scavenge agree with the assumed $P_a$?
   If YES go to (vi)
   If NO put $P_a = P_c$ and go to (v),

v. Return to (i),

vi. STOP.

# 8

## PULSE COMBUSTORS

Pulse combustors are here defined as combustion-driven nonsteady flow devices in which the fuel is reacted in the oxidizer, generally air, intermittently. Furthermore, to qualify for the use of the title pulse combustor it is also assumed, here, that the device is self-aspirating and that this effect is achieved as a consequence of the internal nonsteady flow events. Essentially, then, a pulse combustor is capable of inhaling a flow of air and, provided a supply of fuel is also present, expelling a higher temperature flow of heated fluid at a greater total pressure than that of the entering airstream. Such devices can, therefore, be categorized as gas generators as, for example, was the pressure-exchanger gas generator described in the previous chapter. Normally air-breathing pulse combustors generate a relatively small increase in stagnation pressure, hence they are essentially low-pressure-ratio gas generators and their practical applications are restricted to uses where a small pressure increase is adequate.

In view of the wide variety of pulse-combustor configurations that have been developed as prototypes, some of which have advanced to the market place as commercially available items, the dominant operating principles, and characteristics, common to all pulse combustors will be described first followed by a description of the main conceptual features by means of which most pulse combustors can be classified. This will be followed by a discussion of the application of nonsteady flow theory to the analysis of the performance of pulse combustors. The theoretical analysis of pulse-combustor performance, the principles of which are common to all pulse combustors, invokes many of the concepts introduced in previous chapters with the additional consideration of the combustion, or heat release, process. Finally, practical applications of pulse combustors will be described. Pulse combustors are currently available commercially for several purposes such as commercial, industrial, and domestic heating and also the drying of, for example, foodstuffs. Pulse-combustor fogging units are also available. They are used for distributing, in aerosol form, insecticides, disinfectants, and also for generating smoke screens for military purposes. Many additional uses have also been proposed (Brenchley and Bomelberg, 1984).

### 8.1  Operating Principles

A self-aspirating pulse combustor is normally a tubular or pipelike device, a portion of which, close to one end, is often of enlarged cross section to serve

191

as a combustion zone. The combustion zone is provided with means of ignition such as a spark plug. The fuel, usually in the form of gaseous jets or a spray of liquid, is admitted directly into the combustion zone or into the airstream flowing into the combustion zone or combustion chamber.

In the most basic form combustor an automatic check, or nonreturn, valve is located in the short portion of the pipe communicating with the combustion zone. The check valve is arranged such that flow cannot leave the combustion zone by way of the short pipe, but flow in the opposite direction is restricted as little as possible. The short portion of the pipe, the open end of which communicates with the surroundings, serves, therefore, as an air inlet. The longer portion of the pipe, which is not equipped with a check valve, functions as an exhaust passage to convey products of combustion away from the combustion zone. The pressure rise required to initiate outflow in the exhaust passage, or tailpipe, is due to an increase of pressure occurring in the combustion zone as a result of the discontinuous combustion process. The combustion-generated pressure rise also serves to close the inlet nonreturn valve.

The amplitude of the combustion-generated pressure rise within the combustion zone of pulse combustors varies markedly between units designed for different duties. For pulse combustors intended for domestic heating applications the amplitude of the pressure rise is typically in the region of 10 percent, or less, of the inlet absolute stagnation pressure. For high-duty pulse combustors, such as those intended for propulsion or other power-plant applications, the combustor-generated pressure rise can reach, or exceed, 100 percent of the inlet absolute stagnation pressure. In a conventional pulse combustor combustion is by deflagration and is, therefore, nondetonative. An elemental pulse combustor conforming to the foregoing description is illustrated in Fig. 8.1. The upper portion of Fig. 8.1 shows the combustion process, utilizing air within the combustion zone, causing inlet valve closure and also initiating outflow from the combustion zone into the tailpipe and, subsequently, from the tailpipe outlet.

The inertia of the tailpipe outflow eventually lowers the pressure in the combustion zone below that of the surroundings with the result that the automatic inlet valve reopens and a new charge of air is drawn into the combustion zone. The tailpipe functions, therefore, as a pump to induce air inflow. Vigorous mixing of the fuel entering the combustion zone with the new air charge, and also, very importantly, the residual products of combustion from the previous combustion event, will, for a correctly operating system, ensure that combustion is re-established automatically. The cycle can then be repeated indefinitely. Subsequent induction, mixing, and combustion processes are illustrated, diagrammatically, in the lower portion of Fig. 8.1. In addition to an external source of ignition for start-up, an external air jet, or fan, is, in most cases, also need to create an initial airflow through a pulse combustor prior to self-aspiration being established. The start-up fan or air jet is normally shut off once the pulse combustor is fully operational.

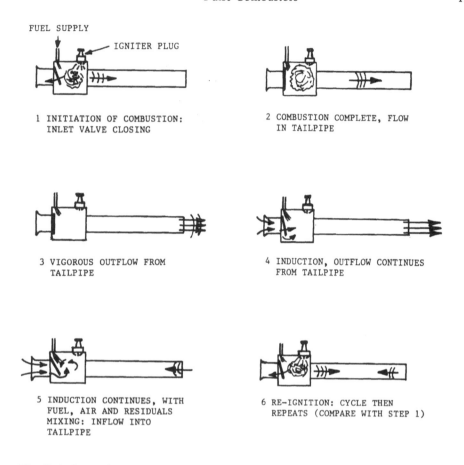

FUEL SUPPLY

IGNITER PLUG

1 INITIATION OF COMBUSTION:
  INLET VALVE CLOSING

2 COMBUSTION COMPLETE, FLOW
  IN TAILPIPE

3 VIGOROUS OUTFLOW FROM
  TAILPIPE

4 INDUCTION, OUTFLOW CONTINUES
  FROM TAILPIPE

5 INDUCTION CONTINUES, WITH
  FUEL, AIR AND RESIDUALS
  MIXING: INFLOW INTO
  TAILPIPE

6 RE-IGNITION: CYCLE THEN
  REPEATS (COMPARE WITH STEP 1)

**Fig. 8–1** Operation of a simple valved pulse combustor (diagrammatic).

In some pulse combustors the fuel is mixed, upstream of the combustion zone with the ingoing entry air, and hence enters the combustion zone in a premixed state. In yet others the mechanical inlet nonreturn valve is omitted. Combustors of the latter type are usually termed valveless or aerodynamically valved (abbreviated to aerovalved) pulse combustors: the design of the air-inlet passages of such combustors will be discussed in more detail later. A crucial aspect of the functioning of pulse combustors concerns the automatic reignition of the charge once the device has been started by means of an external energy source via a spark splug, or possibly a glow plug, installed in the combustion-zone wall.

### 8.1.1  Reignition

It has been shown that charge reignition, once a pulse combustor is operational, occurs due to the presence of residual products of combustion from previous cycles and not, primarily, from any hot surfaces that may be present. It has been demonstrated very clearly, by many workers, that reignition is entirely automatic since the ignition system can be switched off once a pulse combustor is operational. Zhuber-Okrog (1976) showed, by virtue of using a valved pulse combustor with water-cooled combustion chamber walls, that reignition was not occurring due to contact of the fresh charge with high temperature boundary surfaces. Zhuber-Okrog further observed that automatic reignition was associated with the presence of a residual flame from the previous cycle and, while a residual flame was present, use of a multiple-spark-plug ignition system which could be sequenced and phased, electronically, in many operating modes did not influence combustor operation (Zhuber-Okrog, 1968). Zhuber-Okrog's observations were confirmed by subsequent work at the Battelle-Columbus Laboratory (Corliss and Putnam, 1985).

It is virtually a universal finding of all workers that it is essential to mix the incoming charge vigorously with the residual products of combustion present in the combustion zone during the induction process. This serves to reduce, very dramatically, the time required for the reaction process to get under way by virtue of raising the temperature of the incoming charge. It is also an essential process in order to achieve unassisted automatic reignition. Depending upon the expected operational frequency of a pulse combustor such charge preconditioning has proved to be essential in ordering to achieve self-aspirating operation, with automatic reignition, in relatively compact pulse combustors. An inherent characteristic of pulse combustors is a tendency for the emission of oxides in nitrogen ($NO_x$) to be low relative to much other combustion equipment. This beneficial, and significant, effect appears to be the consequence of the rapidity of the combustion process.

### 8.1.2 $NO_x$ emissions

The observation of relatively low $NO_x$ emissions from pulse combustors appear to be due, essentially, to the rapidity of the combustion process, thereby allowing very little time for the formation of oxides of nitrogen, coupled with minimal contact of the products of combustion with residual oxidizer. Peak temperatures lower than those which occur in many alternative combustion situations are in some cases an additional factor. Typically the residence time of the charge in pulse combustors is within the range from about 1 ms for very small units to 10 ms including the time required to complete the combustion process. For many other internal-combustion engines and steady-flow burners typical residence times are up to about one order of magnitude greater. A more detailed account of the influences of residence time, fuel-air ratio, and maximum temperature on the $NO_x$ production of pulse combustors has been given by Putnam et al. (1986). The minimization of $NO_x$ emissions is of considerable importance with respect to the control of pollution of the atmosphere.

## 8.2  Valved Configurations

The earliest pulse combustors were provided with mechanical nonreturn valves; the earliest unit dates from 1906 according to Reader (1977) and other writers (Putnam et al., 1986). The same authors quoted 1909 as the date of the first description of a valveless, or aerovalved, pulse combustor. Despite the proximity of these dates the more obvious and, for some applications more convenient, valved form of pulse combustor was subsequently developed more intensively than the aerovalved variety.

A fundamental feature of valved pulse combustors is the unidirectionality of the through-flow, as shown in Fig. 8.1, since, due to the nature of the device, there is no backflow from the air inlet; such is not the case with most valveless, or aerovalved, configurations. The major problems associated with the design of pulse-combustor inlet valves of the reciprocating mechanical type relate to the requirement of a very low inertia due to the relatively high frequency of pulse combustors, typically within the range from about 50 to 300 Hz, and the need to protect the valves from combustion induced thermal damage. Such problems are greatest with high-duty, high-pressure amplitude, pulse combustors and are further accentuated when the temperature of the inlet air is elevated significantly above that of the atmosphere as can occur, for example, when a pulse combustor is employed as a gas-turbine combustor. For high-duty pulse combustors the automatic valves are usually of the reed type and are fabricated from thin-sheet spring-steel. The spring action of these valves is such that the valves are normally sprung lightly shut. The fully open flow area of inlet valves is normally substantially less than the combustion zone cross-sectional area in order to promote the vigorous mixing referred to previously. Figure 8.2 illustrates two versions of the reed-valve concept. The version on

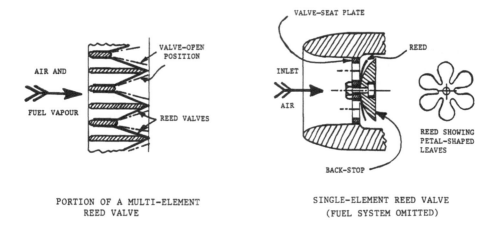

PORTION OF A MULTI-ELEMENT
REED VALVE

SINGLE-ELEMENT REED VALVE
(FUEL SYSTEM OMITTED)

**Fig. 8–2** Two forms of reed valve used on high-duty valved pulse combustors.

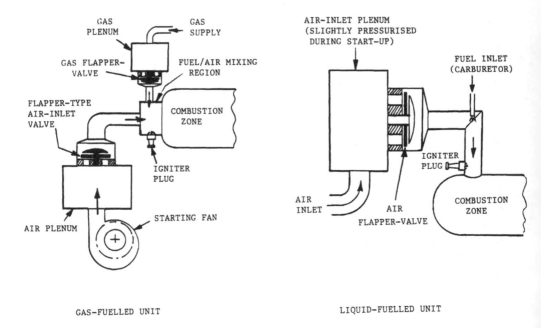

Fig. 8–3 Two types of pulse-combustor flapper valve.

the right, which is employed in some model-aircraft pulse-jet engines (a pulse jet is a pulse combustor used as thrust generator), offers a measure of protection for the reed valve from thermal damage due to the presence of the valve backstop. To date fatigue failure appears to be a major problem with metal reed-type inlet valves. Typically these accumulate from $2 \times 10^5$ to $10^6$ operating cycles per hour of operation.

Valves that move freely, or do not possess any spring action, are termed flapper valves. Nonmetallic flapper valves have been used very successfully on light-duty, natural-gas fueled, pulse combustors such as units used for domestic heating. Here the opportunity exists for distancing the valves from the combustion process, and hence preventing thermal damage from occurring, by providing separate air and gas valves each set back from the combustion chamber in the air and gas inlet pipes, respectively. An arrangement of this type is illustrated, diagrammatically, in the left-hand portion of Fig. 8.3. When a liquid fuel is used in conjunction with a nonmetallic flapper the fuel is either admitted separately downstream of the flapper valve, which therefore controls only the airflow, or in systems where the fuel is premixed with the entering air; thermal protection is provided by means of a flame-trap combustion-quenching device interposed between the valve and the combustion zone. The first arrangement is illustrated diagrammatically in the right-hand portion of Fig. 8.3.

## 8.3  Valveless, or Aerovalved, Configurations

The design of the inlet passages of valveless, or aerovalved, pulse combustors is, from the fluid mechanics viewpoint, more involved than for valved combustors. The dominant advantage of valveless configurations is the complete absence of moving parts with the removal of the attendant risk of their failure. This aspect is of greatest importance for high-duty combustors in which inlet valves experience much more strenuous operating environments than in light-duty units such as domestic heaters, etc.

For many applications the backflow of products of combustion into and, for all but the lightest of duty units, emerging from the air inlet presents a problem. In some applications the backflow can be employed for a useful purpose, for example, entraining additional air that can be directed over the outer surface of the combustor, and hence it is an advantage. The concept selected for the design of the inlet passage of a valveless pulse combustor depends, to a large extent, on the intended use of the backflow.

### 8.3.1  Inlet design concepts

Perhaps the most simple, and obvious, concept for the design of the inlet passage of an aerovalved pulse combustor is to arrange for the inlet passage to, as far as possible, emulate a nonreturn valve. This implies the lowest possible resistance to inflow while offering the greatest possible resistance to backflow. Such a device, which is usually known as a fluid-diode, is by nature always inferior in performance to a mechanical nonreturn valve because, in order to generate a resistance to backflow, backflow must occur. Such a device can, therefore, be regarded as if it were a badly leaking nonreturn valve. The performance of a fluid-diode is usually expressed in terms of the permeability ratio, i.e., the number of inlet-passage-flow dynamic-heads lost during backflow under steady conditions divided by the corresponding loss for inflow. Realistic values of the permeability ratio are typically between about 5 and 20. A fundamental problem of fluid-diodes is that while the backflow can still be utilized for thermal purposes, it is of little value for pumping tasks as a direct consequence of the irreversibilities occurring during backflow. A vortex type fluid-diode, which generates a strong vortex during backflow, but not during inflow, is shown diagrammatically in Fig. 8.4.

A more efficient intake passage design does not involve any attempt to maximize backflow pressure losses but rather is arranged in such a manner that both the inflow and backflow pressure losses are minimized while maximizing, in conjunction with such an inlet, the exploitation of the backflow as an energy source for pumping, pressure-boosting, or thrust generation. In many cases, for example when the pulse combustor is employed as a pulse jet, this involves turning the backflow through 180°. A 180° return bend can be used or the return bend can also be arranged to incorporate the function of a nonsteady-flow ejector, whereby an additional airflow can be entrained that bypasses the pulse combustor. A system of this type is illustrated in Fig. 8.5.

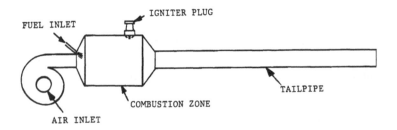

**Fig. 8–4** A vortex type fluid-diode on the inlet passage of a valveless pulse combustor.

For some applications the need for turning the inlet backflow through 180° in conjunction with a low-resistance inlet passage of the type just described can be avoided by employing reward directed inlets. A pulse combustor, with the inlet passage installed in such a manner that it is directed parallel to the tailpipe, has been described by Persechino (1957). An alternative configuration, in which the inlet and tailpipe are arranged, coaxially, at opposite ends of a cylindrical combustion zone, and consequently within which the mixing process should be well organized, incorporates a 180° bend in the pulse-combustor tailpipe. The latter configuration appears to have been originated, in France, by SNECMA and was first used on the SNECMA Ecrevisse (Crayfish) pulse-jet engine (Servanty, 1968).

### 8.3.2   Aerovalve operational characteristics

In addition to the major features of the design of the inlet passages of aerovalved pulse combustors previously discussed, the successful functioning of such passages is subject to the less obvious influences of:

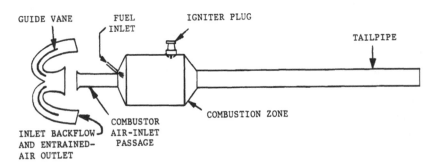

**Fig. 8–5** A valveless pulse combustor with a low-loss inlet and a combined nonsteady flow ejector and 180° bend.

a. inertial confinement,

b. nonuniform cross-sectional area,

c. density nonuniformity.

Each will be considered in turn.

*a. Inertial confinement*

A long inlet of the low-loss type will, because of the inertia of the mass of fluid contained therein, be more effective than a short one in impeding backflow from the combustion zone into the combustor inlet and will, therefore, permit a higher combustion-driven pressure rise to the attained. The induction process following the backflow event has, however, to reinhale into the combustor all the residual products of combustion residing in the inlet passage before air can enter the combustion zone. Hence with regard only to the induction of air into the combustion zone a long inlet is undesirable. These conflicting considerations suggest the likelihood of an optimum inlet length with which a measure of inertial confinement can be achieved, in order to generate strong wave events, without limiting unduly subsequent air induction.

*b. Nonuniform cross-sectional area*

A further mechanism for limiting the backflow from a low-loss inlet passage is to provide a nonuniform cross-sectional area. By tapering the inlet such that it diverges gently toward the combustion zone the inlet somewhat resembles a venturi the throat of which is located immediately downstream of the inlet bellmouth. During the inflow the entering airstream is accelerated up to the throat and thereafter diffuses, with minimum losses, in the diverging section before entering the combustion zone. During backflow the tapered section serves as an efficient nozzle, the effective exit area of which is the intake passage throat area. It is possible, by this means, to in effect provide a larger inlet area than exit area, without introducing any significant irreversibilities, since for inflow the effective passage cross-sectional area is that where the inlet joins the combustion zone, whereas for outflow it is the throat area.

*c. Density nonuniformity*

A fundamental feature of the operation of valveless pulse combustors with low-loss inlets that serves to assist in slowing the backflow of combustion products arises due to the different densities applicable for inflow and for backflow. Although such a density difference is not within the control of the designer, it does serve to inhibit backflow. By way of example it can be shown, on a steady-flow basis, that, to a first approximation, for equal inflow and backflow mass-flow rates the pressure drop for backflow divided by that for inflow is equal to the ratio of the absolute temperature of the backflow divided by that of the inflow.

*8.3.3   Multiple inlets*

As implied in Figs. 8.4 and 8.5, it is customary to arrange for a sudden enlargement at the station where the inlet enters the combustion zone. The area ratio of the enlargement is usually in the region of 3 or greater. The enlargement serves to promote a vigorous mixing process between the incoming airflow, the fuel supplied, and the products of combustion remaining in the combustion zone from the previous cycle. The importance of this mixing process was explained previously in Section 8.1.1.

To a first approximation, for prescribed inflow conditions, the time required for the mixing process is directly proportional to a representative path length which, in turn, is directly proportional to the inlet diameter. Thus by providing, say, four inlets arranged in parallel instead of a single inlet the mixing time for a given total inflow area will be reduced by 50 percent since each inlet of the four-inlet configuration is of half the diameter of the single inlet they collectively replaced. This in turn implies that the four-inlet combustor should be operable at twice the frequency of the single-inlet version with the mixing process occupying the same fraction of the cycle duration in both cases. The advantages of high-frequency operation are such that doubling the operational frequency allows, in principle at least, the overall length of the combustor, which is a nonsteady flow system, to be reduced by 50 percent. In practice, the advantage is somewhat less due to the nonscalable influence of the time occupied by the kinetics of the combustion process. Experimental results obtained using various multiplet-inlet valveless pulse combustors have been reported by several workers included Speirs (1989).

## 8.4   Pulsed Combustors

It is becoming commonplace to refer to cyclic, but nonresonant, combustion devices that do not fire intermittently at the system natural frequency as dictated by wave events but usually at some much lower frequency, often controlled by an ignition, fuel injection, or a valve sequence, as pulsed combustors as distinct from pulse combustors. In many cases pulsed combustors employ power driven rotary valves to control either air inflow, combustion product outflow, or both. Indeed mechanically driven rotary inlet valves have also been used, experimentally, on some pulse combustors despite the inherent problem of maintaining coincident phasing between the rotary valve and the frequency-determining wave events.

Because what are here termed pulsed combustors are not pressure-wave dominated devices they will not be discussed further. The inherent risk of confusion arising due to the use of the similar terms pulsed combustor and pulse combustor should be borne in mind especially since some quite recent literature uses the words "pulsed" and "pulse" synonymously.

Another device, not yet included in this discussion, involving combustion-driven acoustic wave phenomena is the Rijke Tube. Because a Rijke Tube is not self-aspirating, and therefore does not function as a gas generator in the manner of a pulse combustor, it does not qualify, in accordance with the definition given earlier, to be identified as a pulse combustor. Neither is the Rijke Tube strictly a pulsed combustor since the operating frequency is wave controlled.

### 8.4.1   Rijke Tube

The device known as the Rijke Tube after the originator of the concept (Rijke, 1859) is, in the simplest manifestation, a vertical tube, open at both ends, with a heat source provided at approximately one quarter of the tube length above the lower, open, end. The heat source creates a natural, upward, convective flow of the air within the tube. It was found by Rijke that the heat source also resulted in the setting up, and maintenance, of a strong acoustic oscillation within the tube. This phenomenon is in accordance with Rayleigh's criterion (Lord Rayleigh, 1945) which, in effect, states that when heat addition takes place in phase with a pressure oscillation, at a location where the pressure fluctuation is of greatest amplitude, i.e., a pressure antinode, amplification of the oscillation will occur. It has been reported by Putnam (1971) that such a circumstance can be arranged quite naturally, and very easily, where the heat source is a combustion process since the combustion intensity will vary, with a suitable system, essentially in phase with the pressure fluctuation.

The requirement for a Rijke Tube to be arranged vertically can be relaxed if forced rather than natural convection is employed to generate the net airflow through the tubular combustor. Much recent work on coal-, and also wood-, fired Rijke Tubes has been completed by Zinn et al. (1982). Zinn and his co-workers showed that quite significant increases in combustion rate resulted from the nonsteadiness of the flow in their Rijke Tube combustor. The theoretical analysis of Rijke Tube combustors is sometimes treated as, basically, an acoustic problem; however, it would appear that the methods advocated for the analysis of pulse-combustor flows could also be applied. This would require that the coupling of the heat release rate to the local pressure intensity be modeled appropriately and suitable initial conditions imposed to initiate the solution.

### 8.5   Theoretical Treatment

The theoretical modeling of pulse combustors requires that it be possible to account for cyclic one-dimensional transient flows, of finite amplitude rather than of infinitesimal amplitude such as prevail in purely acoustical systems, in ducts often of nonuniform cross-sectional area. Tapered low-loss inlet passages for aerovalved pulse-combustors were described previously in Section 8.3.2: however, tapered tailpipes are also employed in some configurations. In yet other configurations sinuous passages, particularly tailpipes, are used for reasons of compactness or to redirect the flow in a preferred way as, for example, in the case of the Ecrevisse (Section 8.3.1).

As indicated in Chapter 6, Section 6.2.4, no special account need be taken, as a first approximation, of curved passages other than to assign as a substitute for the curved passage an equivalent straight-line length of ducting. Additionally, because the flow passages, particularly the tailpipe, are often of considerable length in relation to their diameters, it is very desirable to include also the influences of wall friction and heat transfer. Any successful theoretical treatment must also account for the nature, and rate of progress, of the combustion process within the combustion zone and the interaction between the combustion zone events and the inlet and tailpipe flows. Differences will exist between versions of the model as applied to aerovalved and valved combustors.

### 8.5.1   Valveless units

Figure 8.6 shows, to scale, the configuration of an experimentally optimized high-duty, single-inlet, aerovalved pulse combustor. It can be seen from Fig. 8.6 that the length-to-average diameter ratio of the tapered tailpipe is in the region of 20:1, for the inlet passage it is approximately 6:1. These values tend to emphasize the previously mentioned desirability of accounting for wall friction and heat transfer. The ends, or boundaries, of the inlet and tailpipe are labeled, respectively, $e_1$, $e_2$ and $e_3$, $e_4$. The ends $e_2$ and $e_3$ bound the enlarged region, of diameter $d$, identified as the combustion chamber. When the combustion chamber is of considerably greater cross-sectional area than the ducts adjoining it, as is the case for the configuration of Fig. 8.6, and also because a major process within the combustion zone involves vigorous mixing, it is reasonable to assume for modeling purposes that, at any instant, the combustion zone contents are of uniform composition and thermodynamic properties and that the internal kinetic energy is negligible. Hence the regions

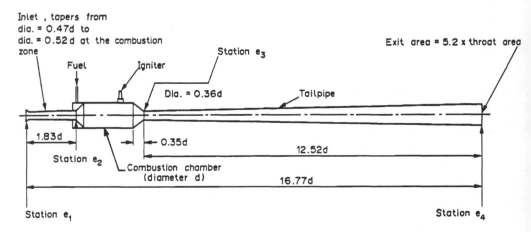

**Fig. 8–6** Proportions of a SNECMA-Lockwood valveless pulse combustor.

of one-dimensional, transient, compressible flow are assumed to be confined to the inlet passage and tailpipe.

The inlet and tailpipe flows can be handled by the methods described in earlier chapters. The appropriate boundary conditions at the ends $e_1$ and $e_4$ communicating with the pulse-combustor surroundings are those applicable to inflow and outflow through fully open ends as presented in Chapter 4. The same boundary conditions can also be applied at the inlet and tailpipe ends, $e_2$ and $e_3$, respectively, communicating with the combustion zone provided the instantaneous combustion zone conditions are substituted for the surrounding conditions applicable to ends $e_1$ and $e_4$. Wall friction and forced-convection heat transfer on the inner walls of the inlet and tailpipe can be estimated, as recommended by Bannister (1964), on the basis of local, instantaneous, parameters using relationships normally applied to steady-flow analysis. Hence the friction factor, $C_F$, is given by:

$$C_F = 0.25 \, \mathrm{Re}^{-0.25} \tag{8.1}$$

and the Nusselt number is given:

$$h_c d_h / k = 0.023 \, \mathrm{Re}^{0.8} \, \mathrm{Pr}^{0.33} \tag{8.2}$$

Equations (8.1) and (8.2) allow the local friction force, $f$, per unit mass and the local heat transfer rate, $q$, per unit mass to be established after invoking other flow properties and the driving temperature difference for heat transfer. It is the normalized forms of $f$ and $q$ that appear ultimately in the flow analysis. Because pulse combustors operate in a cyclic manner, and hence the conditions prevailing at the beginning of the $n$th cycle must also be those applicable at the commencement of the $(n + 1)$th cycle, furthermore, because the influences of both friction and heat transfer are preferably included, it is virtually essential to use a computerized version of the analytical procedure.

The main departure from methods employed in analyzing other cyclic, compressible, nonsteady, flow phenomena hinges upon the need to include the combustion process. Attempts have been made to take into account the various chemical reactions occurring in the combustion of hydrocarbon fuels in order to predict the combustion rate. Such calculations are usually based on Arrhenius type combustion rate equations, one for each species or subspecies involved in the combustion process. This procedure was not found to be satisfactory for the conditions applicable within pulse combustors even for such "simple" fuels as propane ($C_3H_8$). Consequently in an analysis of the operation and performance of a pulse combustor of the type shown in Fig. 8.6 having a combustion zone diameter, $d$, of 76 mm, Olorunmaiye and Kentfield (1989) employed an overall Arrhenius type combustion model as used by Clarke and Craigen (1976) for modeling the combustion process of a pulse combustor. This equation, for propane fuel was:

$$\dot{m}_f''' = K C_{C_3H_8}^{1.5} C_{O_2}^{0.5} \rho^2 e^{-E(\bar{R}T)} \tag{8.3}$$

where $\dot{m}_f'''$ is the mass of fuel consumed per unit volume in a unit of time. The values used for the constants $K$ and $E$ were $4.5 \times 10^6 \, \text{m}^3/\text{kg}\cdot\text{s}$ and 30,000 kJ/kmol, respectively. These values were within the ranges suggested, in the literature, by numerous workers. The mass fraction $C_i$ of the $i$th species is identified in Eq. (8.3) by the appropriate chemical formula.

The assumption was made, for the analysis, that the combustion reaction was complete and that all combustion took place within the combustion zone of the pulse combustor. In practice it is known that very low concentrations of CO and $NO_x$ have been found in the exhausts of propane-fueled pulse combustors. It was further assumed that molecular diffusion into or out of the combustion zone at ends $e_2$ and $e_3$ was negligible. It was thus possible to perform a mass and energy balance within the combustion zone at each step of the calculation and hence establish not only the pressure and temperature therein but also the mass fractions of products and reactants throughout the cycle. The latter results are presented graphically in Fig. 8.7 for a fuel-flow rate representative of about 50 percent of the maximum fuel flow rate of the SNECMA-Lockwood type aerovalved pulse combustor used by Olorunmaiye and Kentfield (1989). The corresponding predicted gas-flow velocities at ends $e_1$, $e_2$, $e_3$, and $e_4$ for the same combustor, at the same fuel flow rate, are shown in Fig. 8.8. Table 8.1 illustrates the progression of the numerical calculations to convergence, in about 12 cycles, from an assumed initial maximum combustion chamber pressure perturbation corresponding to $(P_{max}/P_o) = 1.5$. The experimentally measured values of the maximum and minimum pressure ratios were

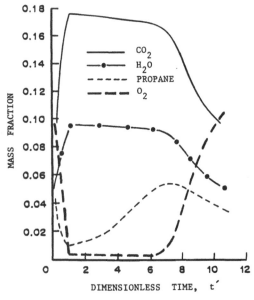

Fig. 8-7 Predicted mass fractions of reactants and products during one cycle of a SNECMA-Lockwood valveless pulse combustor operating at 50 percent of maximum fuel flow-rate.

**Table 8.1** Convergence of Computerized Performance
Predictions for a SNECMA-Lockwood Aerovalved Pulse
Combustor Operating at 50 Percent of Maximum Fuel-Flow Rate

| Cycle Number | $P'_{MAX}$ in Combustion Chamber | $P'_{MIN}$ in Combustion Chamber | Mean Dimensionless Temperature in Combustion Chamber |
|---|---|---|---|
| 1 | 1.500 | 0.801 | 3.84 |
| 2 | 2.240 | 0.761 | 4.88 |
| 3 | 2.227 | 0.786 | 5.18 |
| 4 | 2.126 | 0.790 | 5.41 |
| 5 | 2.073 | 0.785 | 5.44 |
| 6 | 2.073 | 0.783 | 5.35 |
| 7 | 2.102 | 0.783 | 5.11 |
| 8 | 2.166 | 0.774 | 5.17 |
| 9 | 2.236 | 0.777 | 5.17 |
| 10 | 2.228 | 0.775 | 5.21 |
| 11 | 2.221 | 0.776 | 5.23 |
| 12 | 2.220 | 0.777 | 5.22 |
| 13 | 2.226 | 0.776 | 5.22 |
| 15 | 2.230 | 0.776 | 5.23 |
| 15 | 2.230 | 0.776 | 5.23 |
| 16 | 2.229 | 0.776 | 5.23 |

*Note:* Inlet and tailpipe grid spatial dimension, $\Delta X'$, = 25 percent of length of inlet passage.

close to the converged predicted values as were the measured, experimental, and predicted mean gas temperatures.

### 8.5.2 Valved units

The process required for the prediction of the performance of valved units differs from that required for valveless units in that the presence of valves restricts, by positive means, inlet backflow. The most simple of inlet boundary conditions is to assume that inlet valves are either fully open or fully shut: this is equivalent to assuming that inlet valves are of zero mass and hence are inertialess. An application of Newton's Second Law will be required to justify or otherwise such an approximation. It would appear that for conditions typical of many flapper-valved units, such as domestic heaters, valve opening and closing durations can be in the region of 10 percent of the cycle duration and hence, for an accurate analysis, account should be taken of valve inertia possibly also including the influence of nonsteady flow in the passages upstream of valves.

Inlet valve operation can be included in the performance analysis by means of the partly open-end boundary conditions, presented in Chapter 4, in conjunction with application of Newton's Second Law to determine the instantaneous magnitude of valve opening. It may well be, depending upon details of the valve design, that for a fully open valve $\Phi$ is less than unity. The

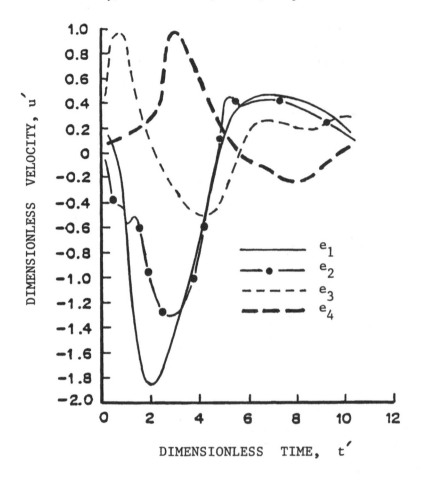

**Fig. 8–8** Predicted velocities at $e_1$, $e_2$, $e_3$, and $e_4$ during one cycle of a SNECMA-Lockwood valveless pulse combustor operating at 50 percent of maximum fuel flow-rate.

majority of pulse combustors currently in commercial production are of the light-duty, or lightly loaded, flapper-valve equipped type. Some highly loaded reed-valve equipped units are also available for model aircraft propulsion.

### 8.6   Practical Applications

Pulse combustors are in commercial production or have recently been tested in the form of prototypes, for a number of practical applications. The first practical applications seem to have been as pulse jets. The German World War II V1 Flying Bomb, or cruise missile in more modern terminology, was propelled by a reed-valve-equipped pulse-jet engine.

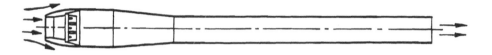

**Fig. 8–9** Reed-valve equipped V1 pulse jet (diagrammatic).

### 8.6.1 Direct propulsion

While the pulse-jet engine of the V1 was successfully applied to the propulsion of the expendable V1 Flying Bomb it was found that the reed-valve life was very short thereby precluding operation over extended periods. The development of the V1 pulse jet, illustrated diagrammatically in Fig. 8.9, was followed, soon after the end of the Second World War, by the production of similar, but much smaller, pulse jets for the propulsion of model aircraft. Most engines of the latter kind were equipped with the type of reed valve shown in the right-hand portion of Fig. 8.2. The short valve life, also experienced with the model aircraft units, is not a serious problem since the valve reed is easily replaced between flights.

Also during the postwar period the development of valveless, or aerovalved, pulse jets was revived in the United States and, particularly energetically, in France. The French work led to the development of several configurations similar to that shown in Fig. 8.6 and also the "U" shaped Ecrivesse referred to previously in Section 8.3.1. For use as a pulse jet the inlet backflow from a configuration of the kind shown in Fig. 8.6 has to be turned through an angle of 180°. This can be accomplished, as in the SNECMA Escopette engine, by means of a return bend. When the return bend is made of variable cross-sectional area, in which a contraction is followed by a gradual enlargement of cross section, the device functions not only as a return bend but also as a nonsteady flow ejector that also augments the thrust generated by the inlet backflow, in addition to reversing the flow direction. A double outflow device of this type is illustrated in Fig. 8.5.

The work at SNECMA led to the discovery that a tapered, divergent, tailpipe was helpful in increasing the effectiveness of the tailpipe as a gas pump for highly loaded aerovalved pulse combustors such as those used for propulsion applications. The SNECMA program resulted in the achievement of maximum specific static thrusts in the region of 34 $kN/m^2$ (700 $lb_f/ft^2$), without the use of thrust augmenters and with ejector-type thrust augmenters suitable only for low forward speed operation up to twice these values, based on the combustion zone cross-sectional area for engines of 0.25 m combustion zone diameter. It was found that for engines of only 0.10 m diameter specific thrusts were reduced to about 50 percent of the above values.

The specific fuel consumption of unaugmented pulse jets at static conditions is very high being typically two to three times that of a nonafterburning turbojet.

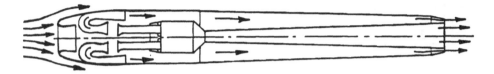

**Fig. 8–10** Proposed four-inlet, aerovalved, ducted, pulse jet for subsonic applications (diagrammatic; to scale).

It appears to be this problem, coupled with the relatively low thrust per unit of frontal area, that this prevented the use of pulse jets for aircraft propulsion. Nevertheless, for short duration flights of small expendable vehicles, such as target drones, the aerovalved pulse jet may yet prove to be attractive due to a low first cost, lightweight, and a hight inherent reliability. Figure 8.10 depicts a recent design for a small ducted pulse jet, for low-cost flight vehicles, based on a four-inlet aerovalved pulse combustor developed by Speirs at the University of Calgary (Speirs, 1989).

### 8.6.2  Pressure-gain combustion

Another application for high-duty, or highly loaded, pulse combustors is as pressure-increasing, or pressure-gain, combustors for thermal plants. The first suggestion of such an application appears to be due to Reynst (Thring, 1961). Reynst proposed, during the Second World War, that pulse combustors be substituted for the conventional steady-flow combustors of gas turbines with the aim of achieving a combustion-driven pressure gain and hence increasing the output of the power turbine or the thrust generated in the case of a turbojet. The concept is illustrated on the temperature-entropy plane in Fig. 8.11. Yet earlier, in 1906, Esnault-Pelterie proposed a simple gas turbine, without a compressor, in which the efflux from a pair of valved pulse combustors impinged on a turbine (Reader, 1977).

   Several attempts have been made, from time to time, to implement Reynst's concept that was based on aerovalved pulse combustors. Notable efforts were made by the French aeroengine manufacturer SNECMA and also by Porter (1958). In neither case were the units actually installed on, and operated in conjunction with, a gas turbine. Recently, after much preliminary work, a valveless pulse, pressure gain, combustor was built, and installed on a very small gas turbine, at the University of Calgary. The maximum stagnation pressure gain achieved experimentally equaled 4 percent of the gas-turbine-compressor absolute delivery pressure [Kentfield and Fernandes, 1990(a)]. It has been shown that due to the improved performance achievable with large pulse combustors relative to those of small size, pressure gains in the region of 6 to 7 percent of the compressor delivery pressure can be expected from similar equipment to that tested at the University of Calgary but of much greater throughput.

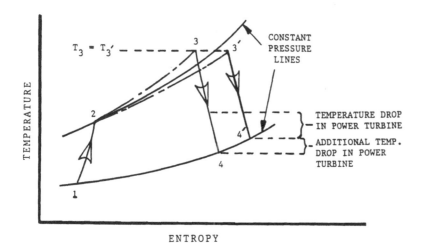

TEMPERATURE

ENTROPY

**Fig. 8–11** Simple shaft-power gas-turbine cycles on the temperature ∿ entropy plane. Cycle 1 2 3' 4' with combustor pressure loss. Cycle 1 2 3 4 with combustor pressure gain.

The benefits obtainable from combustion-driven pressure gain in terms of improved gas-turbine performance depend, to a great extent, on the details of the cycle to which it is applied. Typically about a one to two percent increase in net power output, with a corresponding reduction in specific fuel consumption, can be expected for each one percent reduction in combustion-pressure loss or increase in pressure gain. A quantitative evaluation of the benefits of pressure gain, based on an elementary thermodynamics analysis, is presented in Fig. 8.12. The University of Calgary reverse-flow, prototype, pulse, pressure-gain combustor is illustrated diagrammatically in Fig. 8.13. It should be noted that not only the pulse combustor but also the backflow passage are each of tuned length for optimal functioning of the system.

### 8.6.3   High intensity heating

Highly loaded pulse combustors similar to those used as pulse jets or as pressure-gain combustors can, of course, be used as high intensity heaters. A very significant problem with such applications is the very loud noise generated by pulse combustors of this type due to their very vigorous pressure wave action. The attractiveness of pulse combustors as heaters stems from the high rates of heat transfer achievable due to the flow unsteadiness, as demonstrated by Hanby (1969) and other workers, and also the inherent pumping action permitting products of combustion to be forced through compact heat exchangers, etc.

The noise problems of highly loaded pulse combustors appear to render them quite ill suited for domestic heating purposes; however, they can,

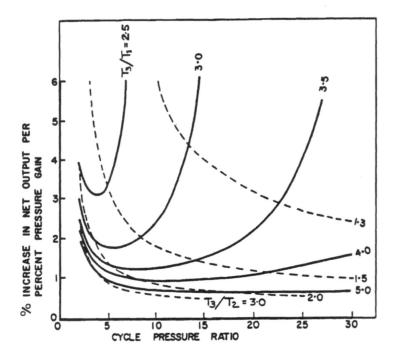

**Fig. 8–12** Influence of combustion-driven pressure gain (simple cycle, shaft power units; see Fig. 8.11 for notation). $\eta_C = \eta_T = 0.85$; $\gamma_C = 1.40$, $\gamma_T = 1.34$; $C_{PC}$ = standard air valve, $C_{PT}/C_{PC} = 1.125$; $\dot{m}_C = \dot{m}_T$.

conceivably, be applied to some industrial or commercial heating tasks especially if noise-control measures are adopted. One way of achieving a measure of noise control is to couple two, or more, combustors to operate sequentially. Briffa and Romaine (1972) showed that very significant noise reductions were achieved with a pair of gas fueled, valved, pulse combustors gas dynamically coupled to operate in antiphase. Later Sran and Kentfield (1982) showed that gas dynamic coupling of either the tailpipes or the inlets, or both, of a pair of aerovalved pulse combustors ensured antiphase operation of units of that type. A problem observed by Sran and Kentfield was, however, a reduction of pumping effectiveness with the coupled units compared with that when uncoupled.

Yet another approach to minimizing the noise of a highly loaded pulse combustor incorporated in a heating plant is to use a single, turbocharged, pulse combustor. The turbocharging simulates the gas-turbine pressure-gain combustor situation described in the previous section. The work on pressure-gain combustion at the University of Calgary demonstrated a dramatic reduction in noise level, compared with that of the isolated combustor, when the combustor was operated installed on a gas turbine. It appears that the

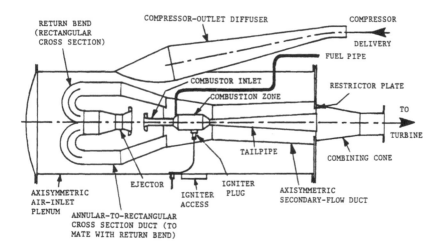

RETURN BEND
(RECTANGULAR
CROSS SECTION)

COMPRESSOR-OUTLET DIFFUSER

COMPRESSOR

DELIVERY

FUEL PIPE

COMBUSTOR INLET

COMBUSTION ZONE

RESTRICTOR PLATE

TO

TURBINE

TAILPIPE

COMBINING CONE

EJECTOR

IGNITER
PLUG

AXISYMMETRIC
AIR-INLET
PLENUM

IGNITER
ACCESS

AXISYMMETRIC
SECONDARY-FLOW DUCT

ANNULAR-TO-RECTANGULAR
CROSS SECTION DUCT (TO
MATE WITH RETURN BEND)

**Fig. 8–13** Configuration of University of Calgary prototype pulse, pressure gain, combustor (diagrammatic).

turbine successfully removes most of the energy from the pulses emitted by the pulse combustor. Analogous observations have been made previously by other workers in connection with noise reductions resulting from the application of turbocharging to reciprocating internal-combustion engines.

In very particular circumstances highly loaded pulse combustors can be used for direct-heating duties without special provisions for noise control. Such an application relates to the rapid prewarming of engines and other equipment used in cold climates. In such cases portability and lightweight preclude the use of noise-control devices since heaters of this type are preferentially hand-held. A prototype hand-held, gasoline fueled, heater of approximately 150 kW thermal output was produced, at the University of Calgary, as a byproduct of some pulse-combustor research (Kentfield, 1977). This device generated an energetic flow of mixed products of combustion and entrained air to impinge on the object to be heated. To minimize the risk of damaging the target the mixed stream temperature was restricted to approximately 200°C. The mass of the apparatus, with sufficient fuel for 15 minutes operation, was approximately 17 kg.

Figure 8.14 shows a "Swingfire" commercially produced, flapper-valve equipped, heater, also gasoline fueled, the simplest version of which is intended for impingement heating although an alternative, interchangeable, turbine driven fan heater, shown in the lower portion of Fig. 8.14, allows the device to function as a warm-air heater with a separate discharge of the pulse-combustor exhaust. The warm-air heater version of the Swingfire unit although intended for heating cabins, vehicle interiors, etc., is generally similar in function to pulse combustors intended for domestic air-heating applications.

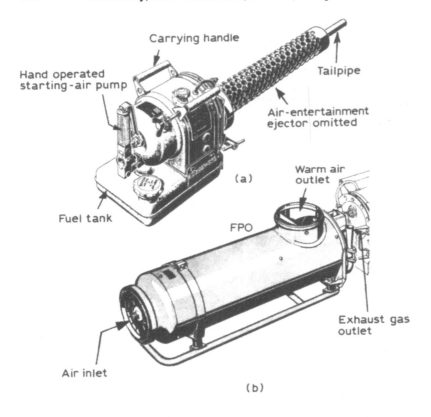

**Fig. 8–14** Eberspächer 'Swingfire' portable pulse-combustor heater. (a) Basic unit, (b) Turboheater (alternative to air ejector) (courtesy J. Eberspächer Gmbh).

### 8.6.4  Domestic and commercial heating

Pulsed combustors, mostly natural-gas fueled, have been applied to domestic and in some cases commercial building, heating systems and units are in production for both air and water heating applications. Although, for domestic units, the trough-to-peak combustion-zone pressure-amplitude are normally low, about 0.07 of the absolute pressure of the surroundings, in order to simplify the problems of noise control very high energy conversion efficiencies are achieved. For household air heating systems the average energy conversion efficiency, usually termed the Seasonally Adjusted Efficiency, is typically in the region of 94–96 percent. Such high efficiencies are possible because the final heat exchange surface is sufficiently cool to condense much of the water vapor contained in the products of combustion. This situation allows the exhaust pipe carrying the products of combustion out of the building to be made from PVC tubing without risk of it melting!

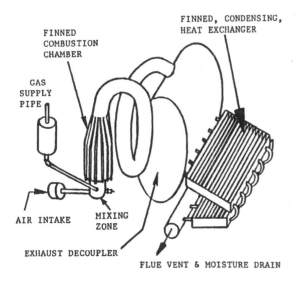

FINNED, CONDENSING, HEAT EXCHANGER

FINNED COMBUSTION CHAMBER

GAS SUPPLY PIPE

AIR INTAKE

MIXING ZONE

EXHAUST DECOUPLER

FLUE VENT & MOISTURE DRAIN

**Fig. 8–15** A Lennox domestic, pulse-combustor, warm-air, house heater (diagrammatic).

Figure 8.15 shows, diagrammatically, the main components of pulse-combustor domestic air heaters manufactured by the Lennox company. For clarity the box-like containment is omitted as is an electric-motor-driven fan for circulating the air to be heated over the air side of the condensing heat exchanger, over the pulse combustor and, subsequently, through the building. The small electrically driven blower, which provides air at a slight overpressure to the pulse-combustor air inlet during start-up, is also omitted. Note the similarity of the air-inlet, and gas, flapper-valve arrangements to that shown in the left-hand portion of Fig. 8.3.

Similar technology has also been applied to domestic water heating. With water heating applications slightly lower average energy conversion efficiencies are achieved than for air heating, about 90–94 percent, due to the temperature of the water in which the pulse combustor is immersed usually being too high to allow very extensive condensation of the water vapor in the combustion products. Figure 8.16 shows, again diagrammatically, a commercially produced domestic water heater, known as the Hydro-Pulse, manufactured by Hydrotherm. Both the inlet valve arrangements and the exhaust decoupler have, for clarity, been omitted from the diagram. The use of multiple tailpipes is worthy of note: these serve to maximize, for a prescribed tailpipe length, the heating surface area available.

Technology in a general sense similar to that of the Hydro-Pulse unit has also been applied, notably by Forbes Engineering, to much larger water heating systems suitable for commercial building, and industrial, uses. Pulse combustors have also been applied to steam-raising boilers.

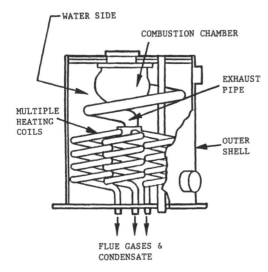

**Fig. 8–16** A Hydrotherm "Hydropulse" pulse-combustor domestic water heater (diagrammatic).

### 8.6.5 *Drying*

Pulse combustors, often of the high-duty, highly loaded, aerovalved type, have been applied to drying applications for human and animal foodstuffs and also for drying substances that would otherwise be waste products in order to create, for example, foods for livestock. The main merit of pulse combustors for drying applications, particularly for drying slurries, lies in the relative motion possible between particles of the substance being dried and the enveloping stream of heated air and products of combustion. Because, almost inevitably, the density of the substance being dried is much greater than that of the gas within which it is immersed, and in which it is being carried, the wave-generated rapid perturbations in the magnitude of the gas velocity are the main contributors to the relative motion between the moist, gas borne, particles and the gas stream. This relative motion is particularly helpful in stripping away, from the particles, the local clouds of vapor in which they would otherwise be enveloped, thereby expediting the drying process.

When drying a slurry with a pulse combustor it is fairly usual to spray the slurry into the tailpipe; most of the drying process occurs downstream of the tailpipe exit. When nonsprayable materials are to be dried, for example crops of various kinds, these are usually placed on mesh racks through, and over, which the efflux of the pulse combustor passes. While some pulse-combustor driers are portable units, most large systems are permanent installations well insulated with respect to noise emissions. Catchment devices, for example exhaust bag houses, are often provided to restrict the emission of very fine particulates. For drying foodstuffs it is particularly important that an inherently clean fuel, for example sulphur-free natural gas, be used to minimize the risk of contamination.

### 8.6.6  Fogging

Pulse-combustor fogging units are always of the portable type. The larger units are usually vehicle mounted, the smaller ones are normally hand held. Usually the pulse combustors are gasoline fueled with flapper-valve-controlled air inlets. The fogging liquid is normally introduced into the pulse-combustor tailpipe, often close to the tailpipe exit, much in the manner of the slurry injection in pulse-combustor driers.

Because of the compact nature of hand-held foggers both the fuel tank and the solution tank containing the fogging liquid are normally built integrally into the unit. The solution tank in larger, usually vehicle-mounted, foggers is often a separate container such as the container within which the solution is supplied. Some fogging units can also be equipped with an alternative flame-thrower head to allow the device to be used for weed burning, etc.

Functionally fogging units differ from other pulse combustors only in the provision of a means for supplying fogging liquid into the tailpipe. Presumably any attempt to predict, in an accurate manner, the performance of a fogging type pulse combustor, or for that matter a slurry drier in which the slurry is also introduced within the pulse combustor, should take into account the consequences of the liquid addition in complicating the fluid mechanics, heat transfer, and thermodynamics of the device.

A small hand-held pulse-combustor fogger is illustrated in Fig. 8.17. This unit, manufactured by Motan Gmbh, is equipped with a hand-operated starting-air pump built into the carrying handle. The ignition, for start-up, is by means of a built-in trembler coil unit energized from flashlight batteries.

### 8.7  Bibliography

A bibliography is included of papers, reports, patents, theses, etc., relating, specifically, to pulse combustion: the entries are listed chronologically. It is intended that the inclusion of a bibliography will be of help to researchers and others, by indicating to them the magnitude, range, and extent of the material available. The contributions of Abbott A. Putnam in assisting with the preparation of the Bibliography are gratefully acknowledged. Works referenced in this Chapter are listed in the Reference section at the end of the book: those that are concerned specifically with pulse combustors can also be found in the Bibliography of Pulse Combustor Literature.

In the interests of brevity papers due to the late F. H. Reynst are not listed individually but are included in a work entitled; *Pulsating Combustion, The Collected Works of F. H. Reynst*. This was published in 1961 under the editorship of M. W. Thring. Similarly, papers presented at each of three international symposia on pulsating combustion are also not listed separately. The first of these symposia was held in 1971 and the proceedings were published,

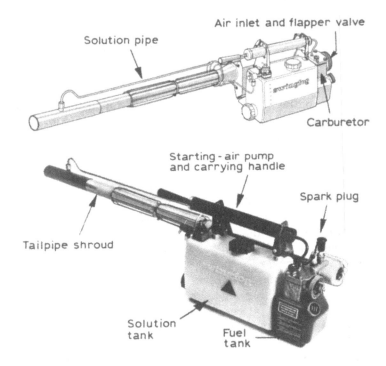

**Fig. 8–17** A "Swingfog" pulse-combustor, hand-held, fogging unit manufactured by Motan Gmbh (courtesy Motan Gmbh).

in that year, under the editorship of D. J. Brown. The second symposium was held in 1979 and the proceedings, which appeared in 1980, were edited by M. J. Clinch. The 1982 symposium, entitled; *Symposium on Pulse Combustion Applications*, was held in 1982 at Atlanta, Georgia. Since this symposium was combined with another symposium on other aspects of combustion only Vol. 1 of the full set of proceedings relates to pulse combustion. Approximately twenty papers on pulsating combustion were presented at each of the three symposia.

It can be seen from the Bibliography, which covers the period from the end of the Second World War to early 1990, that for about the first twenty-year period emphasis was on the direct application of pulse combustors, in the form of pulse-jet engines, for flight propulsion including the tip-drive of helicopter rotors. Latterly emphasis has changed to other applications, as described in Sections 8.6.2 to 8.6.6 inclusive, and more fundamental studies.

Bibliography of Pulse-Combustor Literature

| Year | Author, Authors, or Lead Author | Title and Publication |
|------|-------------------------------|----------------------|
| 1945 | Bailey, N. P. and Wilson, H. A. | The Intermittent Jet Engine. U.S. Government Printing Office, Washington, D.C. |
| 1945 | Everett, H. A. | Preliminary Study of the Possibilities of Utilizing Combustion Chambers of the Intermittent (or Aero-Pulse) Type in Conjunction with Gas Turbine Power Plants. Avco Lycoming Div., Williamsport, Pa. Report No. 853. |
| 1945 | Diedrich, G. | The Aero-Resonator Power Plant of the V-1 Flying Bomb. (in German: Princeton University translation 1948). AD-495754. |
| 1945 | Bressman, J. R. | Effect of Low-Loss Air Valve on Performance of 22 Inch Diameter Pulse-Jet Engine. NACA Wartime Report No. E-279. |
| 1946 | Macdonald, J. K. L. | A Gas Dynamical Formulation for Waves and Combustion in Pulse Jets. Report No. 151, Applied Mathematics Group, New York University. |
| 1946 | Goddard, R. H. | U.S. Patent No. 2,409,036. |
| 1947 | Edelman, L. B. | "Pulsating Jet Engine—Its Evolution and Future Prospects," Transactions of SAE, Vol. 1, No. 2, pp. 204–216. |
| 1947 | Wilder, J. G. | Preliminary Experimentation on the Cornell Aeronautical Laboratory 6″ and 4″ Pulse Jets. Report No. TM CAL 6DD 420 A6, James Forrestal Research Center, Princton University. |
| 1947 | Schultz-Grunow, F. | Gas Dynamic Investigation of the Pulse-Jet Tube, NACA TM-1131. |
| 1948 | Torda, P. et al. | Compressible Flow Through Reed Valves for Pulse Jet Engines. Project SQUID Technical Report No. 9, Polytechnic Institute of Brooklyn. |
| 1948 | Zipkin, M. A. et al. | Analytical and Experimental Performance of an Explosion Cycle Combustion Chamber for a Jet-Propulsion Engine. NACA TN 1072. |
| 1948 | Reissner, H. J. and Torda, P. | Project Squid: Final Report Phase 1, Polytechnic Institute of Brooklyn. |
| 1948 | Logan, J. G. | Suggested Forms of Air Duct Motors Utilizing Intermittent Combustion. Report CAL-16-M. Cornell Aeronautical Laboratory, Buffalo, N.Y. |

| 1949 | Oppenheim, A. K. | Research and Development of Impulsive Ducts in Germany. British Intelligence Objectives Sub-Committee Report No. 1777, Technical Information and Documents Unit, London. |
|---|---|---|
| 1949 | Schmidt, J. H. et al. | Model XPJ 40-MD2 Valveless Pulse Jet Engine Development. McDonnell Aircraft Corp., Report 1509. |
| 1949 | Segreto, J. J. and Striger, K. | Pulse-Jet Ground Heating Devices, Air Material Command Royal Canadian Air Force. Report No. MCREXE-657-186F. |
| 1950 | Goddard, R. H. | U.S. Patent No. 2,515,644. |
| 1950 | Kollsman, P. | U.S. Patent No. 2,523,379. |
| 1950 | Dunbar, J. Y. | U.S. Patent No. 2,525,782. |
| 1951 | Bodine, A. G. | U.S. Patent No. 2,543,758. |
| 1951 | Bodine, A. G. | U.S. Patent No. 2,546,966. |
| 1951 | Bohanon, H. R. | U.S. Patent No. 2,574,460. |
| 1951 | Logan, J. G. | Summary Report on Valveless Pulsejet Investigation. Technical Memorandum CAL-42, Cornell Aeronautical Laboratory, Buffalo, N.Y. |
| 1951 | Torda, P. | "Approximate Theory of Compressible Flow Through Reed Valves for Pulse Jets," *Proceedings of the First Midwestern Conference on Fluid Dynamics*, Ann Arbor, Mich. |
| 1951 | Rudinger, G. | On the Performance Analysis of the Ducted Pulsejet. Project SQUID Report T.M. No. CAL-36. Cornell Aeronautical Laboratory, Buffalo, N.Y. |
| 1952 | Salmon, B. | Examen de la Flamme dans la Chambre de Combustion du Pulso Escopette. SNECMA, Note ES XIII-10. |
| 1952 | Salmon, B. | Pressions Instantanées dans la Trompe de Dilution du Pulso 5158 Avec Capacité Arriéré. SNECMA, Note ES XIII-14. |
| 1952 | Salmon, B. | Examen de la Flamme dans la Chambre de Combustion du Pulso 509. SNECMA, Note ES XIII-17. |
| 1952 | Salmon, B. | Examen de la Flamme dans la Chambre de Combustion du Pulso, 5158, SNECMA, Note ES XIII-18. |
| 1952 | Vigne, H. | Compte Rendu d'Essais de Post-Combustion en Aval du Pulso 5158. SNECMA, Note ES II-68. |

| 1952 | Schmitt, H. | Examen de al Flamme dan un Pulso a Ecoulment Bidimensional. SNECMA, Note ES XIII-24. |
|------|------------|-------------------------------------------------------------------------------------|
| 1952 | Temte, G. H. | U.S. Patent No. 2,580,484. |
| 1952 | Bodine, A. G. | U.S. Patent No. 2,581,902. |
| 1952 | Black, L. E. | U.S. Patent No. 2,587,100. |
| 1952 | Bauger, L. A. J. | U.S. Patent No. 2,593,523. |
| 1952 | Tenney, W. L. et al. | U.S. Patent No. 2,599,209. |
| 1952 | Yen, S. M. | "Gas Dynamic Investigation of a Valveless Pulse-Jet Tube," *Proceedings of the Second Midwestern Conference on Fluid Dynamics*, Ohio State University. |
| 1952 | Tenney, W. L. et al. | U.S. Patent No. 2,609,660. |
| 1952 | Tenney, W. L. | U.S. Patent No. 2,612,722. |
| 1952 | Tenney, W. L. et al. | U.S. Patent No. 2,612,748. |
| 1952 | Tenney, W. L. et al. | U.S. Patent No. 2,612,749. |
| 1952 | Tenney, W. L. et al. | U.S. Patent No. 2,612,955. |
| 1953 | Traenckner, K. | "Pulverized-Coal Gasification Ruhrgas Processes," *Trans. ASME*, Vol. 75, pp. 1095–1097. |
| 1953 | Porter, C. D. | Cornell Valveless Pulsejets Tests and Further Developments. Naval Research Laboratory, Washington, D.C., Report No. NRL-4186. |
| 1953 | Lockwood, R. M. | Pulsejet Ejectors. Dissertation for the professional degree of Mechanical Engineer. Oregon State University. |
| 1953 | Tenney, W. L. et al. | U.S. Patent No. 2,633,703. |
| 1953 | Kamm, W. I. E. et al. | U.S. Patent No. 2,643,107. |
| 1953 | Myers, E. B. | U.S. Patent No. 2,657,708. |
| 1953 | Bertin, J. et al. | U.S. Patent No. 2, "Le Pulso-Reacteur SNECMA L'Escopette," *Interavia*, Vol. 8, No. 6, pp. 343–347. |
| 1954 | Tenney, W. L. | U.S. Patent No. 2,675,670. |
| 1954 | Kallen, H. | "Pulsating Combustion: An Old Idea May Give Tomorrow's Boilers a New Look," *Power*, Vol. 98, Aug. |
| 1954 | Schiefer, R. B. | Pulsejet, Test of Naval Research Laboratory Six-Inch Diameter Pulsejet. General Electric Co., Report No. DF 54 TG353. |
| 1954 | Lockwood, R. M. and Thomas, C. H. | Development and Sea-Level Static Tests of American Helicopter Five-Inch Diameter |

|       |                                    | Pulse-Jets. Fairchild Electronics Div., Fairchild Engine and Airplane Corp., Costa Mesa, CA. Report No. 1949 B 401. |
|-------|------------------------------------|-----|
| 1954  | Demartinis, S. A.                  | Structural Development of High Performance Inlet Valves for Use on Rotary-Wing Tip-Mounted Pulse-Jet Engines. Fairchild Electronics Div., Fairchild Engine and Airplane Corp., Costa Mesa, CA. Report No. 1949 B 412. |
| 1954  | Lockwood, R. M. and McJones, R. W. | Combustion Chamber Average Pressure as a Pulsejet Internal Performance Indicator. Fairchild Electronics Div., Fairchild Engine and Airplane Corp., Costa Mesa, CA. Report No. 1949 B 423. |
| 1954  | Lockwood, R. M.                    | Primary Data Presentation From Exploratory Altitude Tests of AJ-5.0 Pulsejet in WADC Dynamometer No. 24. Fairchild Electronics Div., Fairchild Engine and Airplane Corp., Costa Mesa, CA. Report No. 1949 B 402. |
| 1954  | Hertzberg, A. and Russo, A.        | Experimental Investigation of a Resonant Wave Engine. Cornell Aeronautical Laboratory Inc. Report No. DD 799 A2. |
| 1954  | Lockwood, R. M. and Dunn, R. E.    | High Altitude Development and Performance Trends of Rotary-Wing Tip-Mounted Pulse-Jet-Engines. Fairchild Electronics Div., Fairchild Engine and Airplane Corp., Costa Mesa, CA. Report No. 172 D 2. |
| 1954  | Lockwood, R. M.                    | Ducted Pulsejets for High Subsonic, High Altitude, Propulsion. Project SQUID Technical Report PR-55-C, Princeton University. |
| 1954  | Bertin, J.                         | Quelques Propriétés de la Combustion Pulsatoire: le Pulso-Réactoire Action des Carburats Deopes. AGARD. Selected Combustion Problems, pp. 490–505. |
| 1954  | Cunningham, W. R.                  | Development of High Performance Rotary-Wing Tip Mounted Pulsejet Engine Components. Final Report No. 172-D-3 (Nov.). American Helicopter Company. |
| 1954  | Foa, J. V.                         | The R.P.I. Wave Engine: Summary Report to Oct. 31, 54. Technical Report TR AE 5409 for the Office of Naval Research. Dept. of Aeronautical Engineering, Rensselaer Polytechnic Institute, Troy, New York. |
| 1955  | Lustig, L.                         | U.S. Patent No. 2,703,565. |
| 1955  | Huber, L. et al.                   | U.S. Patent No. 2,708,926. |

| 1955 | Frank, P. A. et al. | U.S. Patent No. 2,710,524. |
| 1955 | Tenney, W. L. et al. | U.S. Patent No. 2,715,390. |
| 1955 | Lafferentz, B. et al. | U.S. Patent No. 2,715,436. |
| 1955 | Huber, L. | U.S. Patent No. 2,171,637. |
| 1955 | Haag, F. A. et al. | U.S. Patent No. 2,719,580. |
| 1955 | Porter, C. D. and Persechino, M. A. | Velocity Tests on NRL Valve Pulsejet and Whirling Loss Evaluation. U.S. NRL Report No. 4621. |
| 1956 | Tenney, W. L. et al. | U.S. Patent No. 2,738,334. |
| 1956 | Kadosch, M. et al. | U.S. Patent No. 2,738,646. |
| 1956 | Carpenter, et al. | Investigation of the Propulsive Characteristics of a Helicopter Type Pulsejet Engine Over a Range of Mach Numbers. NACA TN-3625. |
| 1956 | Tenney, W. L. et al. | U.S. Patent No. 2,768,031. |
| 1956 | Russo, A. L. et al. | The Investigation of a Resonant Wave Engine. Cornell Aeronautical Laboratory Inc. Report No. DD 799 A 3. |
| 1956 | Tenney, W. L. | U.S. Patent No. 2,770,226. |
| 1957 | Bodine, A. G. | U.S. Patent No. 2,796,734. |
| 1957 | Bodine, A. G. | U.S. Patent No. 2,796, 735. |
| 1957 | Gosslau, F. | Development of the V1 Pulse-Jet. Agardograph No. 20, pp. 400–418. Appelhans, Brunswick. |
| 1957 | Tenney, W. L. et al. | U.S. Patent No. 2,799,137. |
| 1957 | Kadosch, M. et al. | U.S. Patent No. 2,801,515. |
| 1957 | Dini, D. | "Applicazione Della Teoria Della Corrente Unidimensionale Allo Studio Del Moto Non Stazionario Del Fluido In Pulsogetti Ruotanti," *Aerotecnica*, Vol. XXXVII, No. 3. |
| 1957 | Wilman, S. | U.S. Patent No. 2,805,545. |
| 1957 | Bertin, J. and Salmon, B. | *Combustion Research and Reviews*, Butterworth, London, pp. 122–123. |
| 1957 | Lockwood, R. M. | A New Concept of Resonant Combustion. Hiller Aircraft Corporation. Report No. ARD-129. |
| 1957 | Logan, J. G. | Summary Report on Valveless Pulsejet Investigation. Technical Memorandum CAL-42, Cornell Aeronautical Laboratory, Buffalo, N.Y. |

| 1957 | Persechino, M. A. | Valveless Pulsejet De-Icer Application. U.S. NRL Report No. 5024. |
|---|---|---|
| 1957 | Le Foll, J. et al. | U.S. Patent No. 2,812,635. |
| 1957 | Meulien, H. L. P. et al. | U.S. Patent No. 2,814,930. |
| 1958 | Tenney, W. L. | U.S. Patent No. 2,821,986. |
| 1958 | Ryder, F. A. et al. | U.S. Patent No. 2,834,336. |
| 1958 | Kamm, W. I. E. | U.S. Patent No. 2,839,046. |
| 1958 | Porter, C. D. | "Valveless-Gas-Turbine Combustors with Pressure Gain," ASME Paper No. 58-GTP-11. |
| 1958 | Marks, C. B. et al. | U.S. Patent No. 2,853,995. |
| 1958 | Tenney, W. L. et al. | U.S. Patent No. 2,857,332. |
| 1958 | Zhuber-Okrog, G. | "Principles of Pulsating Combustion Applied to Gas Turbines," D.I.C. dissertation, Dept. of Mech. Engineering, Imperial College, University of London. |
| 1958 | Lockwood, R. M. | Proposal for Development of an Augmented Valveless Pulsejet Lift-Propulsion System. Hiller Aircraft Corporation. Report No. ARD-196. |
| 1958 | Lockwood, R. M. | Proposal for Investigation of the Process of Energy Transfer from an Intermittent Jet to an Ambient Fluid. Hiller Aircraft Corporation. Report No. ARD-199. |
| 1958 | Lockwood, R. M. | Investigation of a Resonant Combustor Concept. Hiller Aircraft Corporation. Report No. ARD-219. |
| 1959 | Lockwood, R. M. | Investigation of the Process of Energy Transfer from an Intermittent Jet to an Ambient Fluid. Hiller Aircraft Corporation. Report No. ARD-238. |
| 1959 | Paris, F. G. et al. | U.S. Patent No. 2,878,790. |
| 1959 | Kitchen, J. A. et al. | U.S. Patent No. 2,898,978. |
| 1959 | Lang, T. G. | U.S. Patent No. 2,903,850. |
| 1959 | Kumm, E. L. | U.S. Patent No. 2,911,957. |
| 1959 | Persechino, M. A. | Experimental Valveless Pulsejet Diesel-Fueled Fog Generator. U.S. NRL Report 5414, Dec. |
| 1959 | Kitchen, J. A. | U.S. Patent No. 2,916,032. |
| 1960 | Terpe, G. R. | U.S. Patent No. 2,925,069. |
| 1960 | Heuer, G. E. and Lockwood, R. M. | Investigation of a Resonant Combustor Concept. Hiller Aircraft Corporation. Report No. ARD-253. |

| 1960 | Lockwood, R. M. et al. | Thrust Augmented Intermittent Jet Lift-Propulsion System. Hiller Aircraft Corporation. Report No. ARD-256. |
| 1960 | Foa, J. V. | *Elements of Flight Propulsion*, Wiley, N.Y. |
| 1960 | Collinson, E. S. | U.S. Patent No. 2,965,079. |
| 1961 | Roginskii, O. G. | "Oscillatory Combustion," *Soviet Phys. Acoust.*, Vol. 7, pp. 131–154. |
| 1961 | Thring, M. W. (Editor) | *Pulsating Combustion. The Collected Works of F. H. Reynst*, Pergamon Press. |
| 1961 | Lockwood, R. M. | Interim Summary Report on Investigation of The Process of Energy Transfer from an Intermittent Jet to Secondary Fluid in an Ejector Type Thrust Augmenter. Hiller Aircraft Corporation. Report No. ARD-286. |
| 1961 | Porter, C. D. et al. | U.S. Patent No. 2,998,705. |
| 1961 | Salgo, L. et al. | U.S. Patent No. 3,005,485. |
| 1962 | Wojcicki, S. | Pulsejet, Jet and Rocket Engines. Wyclownictwo, Ministerstwa. Obrany Narodowej, Poland. |
| 1962 | Lockwood, R. M. | Summary Report on Investigation on Miniature Valveless Pulsejets. Hiller Aircraft Corporation. Report No. ARD-307. |
| 1962 | Linderoth, E. T. | U.S. Patent No. 3,305,413. |
| 1962 | Babkin, Yu. L. | "Studies of Pulsating Combustion of Liquid Propellants," *Transactions of First All-Union Conference on Pulsating Combustion. GIAP* (in Russian). |
| 1962 | Francis, W. E. et al. | A Study of Gas-Fired Pulsating Combustors for Industrial Applications. The Gas Council, London. Research Comm. GC91 (see also Francis et al., 1963). |
| 1962 | Lockwood, R. M. and Patterson, W. G. | Interim Summary Report Covering the Period 1 April 1961 to 30 June 1962 on Investigation of The Process of Energy Transfer From an Intermittent Jet to Secondary Fluid in an Ejector Type Thrust Augmenter. Hiller Aircraft Corporation. Report No. ARD-305. |
| 1963 | Griffiths, J. C. et al. | New or Unusual Burners and Combustion Processes (part 3A Pulse Combustion). A.G.A. Research Bulletin No. 96. |
| 1963 | Lockwood, R. M. | Hiller Pulse Reactor Lift-Propulsion Development Program—Final Report. Hiller Aircraft Corporation. Report No. ARD-308. |

| | | |
|---|---|---|
| 1963 | Marchal, R. and Servanty, P. | Note Sur Le Développement des Pulso-Réacteurs Sans Clapets. CR. ATMA, pp. 611–633 (see also Session Association Technique Maritime et Aéronautique, pp. 1–20). For English translation see CSIR (Pretoria, South Africa), Report N71-31102 of 1970). |
| 1963 | Rydberg, J. A. | U.S. Patent No. 3,091,224. |
| 1963 | —— | Surveillance Test (Environmental) of Generator Smoke, Mechanical, Pulse Jet, ABC-M3A3. Dugway Proving Ground, Utah, Report No. DGP-371. |
| 1963 | Francis, W. E. et al. | "A Study of Gas-Fired Pulsating Combustors for Industrial Applications," *Journal of the Institute of Gas Engineers*, Vol. 3, p. 302 (see also Francis et al., 1962). |
| 1963 | Moen, D. A. | "Investigation of the Performance Characteristics of Two Valveless Pulset Jets Operating in Phase Opposition," MSc dissertation, Univ. of North Dakota, Grand Forks. |
| 1964 | Huber, L. | "Swingfire Operated Heating Units for Commercial Vehicles," *ATZ Technical Review*, No. 2, pp. 31–37. |
| 1964 | Schmidt, P. | "Instationäre Gasdynamische Vorgänge als Grundlage Technischer Verfahrensweisen," *VDI.Z*, No. 11, p. 106. |
| 1964 | Taylor, D. S. | "The Pulsating Combustion of Pulverized Coal," PhD dissertation, Univ. of Sheffield. |
| 1964 | Lockwood, R. M. and Sander, H. W. | Investigation of the Process of Energy Transfer from an Intermittent Jet to a Secondary Fluid in an Ejector Type Thrust Augmenter—Interim Summary Report Covering the Period 30 Oct. 1962 to 31 March 1964. Hiller Aircraft Corporation. Report No. APR-64-4. |
| 1964 | Lockwood, R. M. | "Pulse-Reactor Low Cost Lift-Propulsion Engines," AIAA Paper No. 64-172. |
| 1965 | Melenric, J. A. | U.S. Patent No. 3,166,904. |
| 1965 | Haag, F. et al. | U.S. Patent No. 3,169,570. |
| 1965 | Klein, H. C. | U.S. Patent No. 3,175,357. |
| 1965 | Griffiths, J. C. and Niedzwiecki, R. W. | "An Evaluation of Some New or Different Ways to Heat the Home," A.G.A. Research Bulletin No. 101. |

| 1965 | Tharratt, C. E. | "The Propulsive Duct," *Aircraft Engineering*, Nov., pp. 327–337, and Dec., pp. 359–371. |
| 1965 | Babkin, Yu. L. | "Pulsating Combustion Chambers as Furnaces for Steam Boilers," *Tepoénergetica*, Vol. 12, No. 9, pp. 23–27. |
| 1965 | Bodine, A. G. | U.S. Patent No. 3,185,871. |
| 1965 | Melenric, J. A. | U.S. Patent No. 3,188,804. |
| 1965 | Haag, F. | U.S. Patent No. 3,192,986. |
| 1965 | Marchal, R. et al. | U.S. Patent No. 3,194,295. |
| 1965 | Lockwood, R. M. | U.S. Patent No. 3,206,926. |
| 1965 | Aldag, H. W. | "The Pulse-Jet Engine as a Prime Mover," ASME Paper 65-WA/GTP-6. |
| 1965 | Sargent, E. R. | Development of Multiple Pulse Reactor Package VTOL Lift Engine '6 Pack'. Hiller Aircraft Div. of Fairchild Hiller Corporation. Report No. AP/R-65-16. |
| 1966 | Rocketdyne Div. of North American Rockwell Inc. | Investigation, Design and Test of Resonant Combustor Turbine Starters. Final Technical Report No. R-6883. |
| 1966 | Tharratt, C. E. | "The Propulsive Duct," *Aircraft Engineering*, Jan., pp. 23–25. |
| 1966 | Huber, L. | U.S. Patent No. 3,246,842. |
| 1966 | Marchal, R. and Servanty, P. | "Turbines à Gaz à Chambres de Combustion Pulsatoire," *Entropie* (Sept.–Oct.), No. 11, pp. 37–40. |
| 1966 | Pallabazzer, R. | Studio Critico Del Pulsoreattore (Critical Examination of the Pulse-Jet). Monografie Scientifiche e Techniche di Aeronautica. Under the auspices of the Centro Consultivo Studio e Ricerche dell'Aeronautica Militare, No. 27, pp. 1–29. |
| 1966 | Kitchen, J. A. | U.S. Patent No. 3,267,985. |
| 1966 | Hanby, V. I. | "Convective Heat Transfer in a Gas-Fired Pulsating Combustor," *Journal of Engineering for Power*, Vol. 91A, pp. 48–52. |
| 1966 | Ellman, R. et al. | "Adapting a Pulse-Jet Combustion System to Entrained Drying of Lignite," *Proceedings*, 5th International Coal Preparation Congress, pp. 462–476. |
| 1967 | Maersh-Moller, H. | Changing the Noise Spectrum of Pulse Jet Engines. B and K Technical Review No. 4. |
| 1967 | Llobet, A. F. et al. | U.S. Patent No. 3,323,304. |

| 1967 | Muller, J. L. | "The Development of a Resonant Combustion Heater for Drying Applications," *The South African Mechanical Engineer*, Vol. 16, Feb., pp. 137–146. |
|------|--------------|------------------------------------------------------------------------------------------------------------------------------------------------|
| 1967 | Fuller, L. E. and Maddox, J. P. | Experimental Gas Turbine Engine Starter—Interim Report. Rocketdyne Report No. R-7217. |
| 1967 | Porter, J. W. | "Pulsating Combustion of Liquid Fuels in Partially Closed Vessels," *Combustion and Flame*, Vol. 11, pp. 501–510. |
| 1968 | Grebe, J. J. | U.S. Patent No. 3,365,880. |
| 1968 | Hanby, V. I. and Brown, D. J. | "A 50 lb/h Pulsating Combustor for Pulverized Coal," *Journal of the Institute of Fuel*, Vol. 41, pp. 423–426. |
| 1968 | Zhuber-Okrog, G. | Uber die Wirkungsweise von Schmidt-Rohren, Forschungsbericht 68-49, DFVLR. |
| 1968 | Burgner, G. R. | Feasibility of Pulsejets and Intermittent Combustion Devices as Modern Propulsion Power Plants. Naval Weapons Center Report NWC TP 4536. |
| 1968 | Pornsiripongse, V. | "An Examination of the Performance of a Pressure Generating Pulsed Combustor," M.Sc. dissertation, Imperial College, Univ. of London. |
| 1968 | Marchal, R. | "Turbines à Gaz à Chambres de Combustion Pulsatoire," *Entropie*, No. 22, July–Aug., pp. 15–19. |
| 1968 | Servanty, P. | "Réflexions sur la Combustion Pulsatoire," *Entropie*, No. 22, July–Aug., pp. 49–63. |
| 1969 | Hanby, V. I. and Brown, D. J. | The Pulsating Combustion of Pulverized Coal. 3rd All Union Conference on the Combustion of Solid Fuels. Novosibirsk, Siberia. |
| 1969 | Hanby, V. I. | "Convective Heat Transfer in a Gas-Fired Pulsating Combustor," ASME *Journal of Engineering for Power*, Vol. 91, Series A, pp. 48–52. |
| 1969 | Katsnel'son, B. D. et al. | "An Experimental Study of Pulsating Combustion," *Teploénergetica*, Vol. 16, No. 1, pp. 3–6. |
| 1969 | Severyanin, V. S. | "The Combustion of Solid Fuel in a Pulsating Flow," *Teploénergetica*, Vol. 16, No. 1, pp. 6–8. |
| 1969 | Griffiths, J. C. and Weber, E. J. | The Design of Pulse Combustion Burners. A.G.A. Research Bulletin 107. |
| 1969 | Soper, W. G. | Experiments With a 4.5-inch Pulsejet Engine. Naval Weapons Laboratory, Dahlgren, Virginia. Unpublished Report. |

| 1969 | Graber, D. A. | U.S. Patent No. 3,456,441. |
|------|--------------|----------------------------|
| 1969 | Lockwood, R. M. et al. | U.S. Patent No. 3,462,955. |
| 1969 | Brown, E. W. | U.S. Patent No. 3,486,331. |
| 1969 | Hayek, A. E. | "Analysis of a Pulsed Combustor With an Entrained Secondary Flow," M.Sc. dissertation, Imperial College, Univ. of London. |
| 1969 | Fernando, S. W. | "Multiple Inlet Pulsed Combustor," DIC dissertation, Imperial College, Univ. of London. |
| 1969 | Muller, J. L. | Theoretical and Practical Aspects of the Application of Resonant Combustion Chambers in the Gas Turbine. National Mechanical Engineering Research Institute, CSIR, South Africa. Report MEG 831. |
| 1969 | Mosely, P. E. and Porter, J. W. | "Helmholtz Oscillations in Pulsating Combustion Chambers," *Journal of the Acoustical Society of America*, Vol. 46, No. 1, Part 2, pp. 262–266. |
| 1969 | Belter, J. W. et al. | "Operating Experience With Lignite-Fueled Pulse-Jet Engines," ASME Paper No. 69-WA/FU-4. |
| 1970 | Lockwood, R. M. et al. | U.S. Patent No. 3,498,063. |
| 1970 | Meyer, P. O. | U.S. Patent No. 3,503,383. |
| 1970 | Brown, D. J. and Hanby, V. I. | "High Intensity Combustion," Proceedings of the North American Fuel Technology Conference, Ottawa, Canada, Published by Institute of Fuel. |
| 1970 | Cronje, J. S. | "An Improved Ducted Pulsed Combustor," M.Sc. dissertation, Imperial College, Univ. of London. |
| 1970 | Sadiq, S. M. P. | "Development of a Multiple Inlet Pulsed Combustor," M.Sc. dissertation, Imperial College, Univ. of London. |
| 1971 | Binsley, R. L. | Application of a Resonant Combustor to Army Aircraft Starting. Final Technical Report, 1 Sept. 69–30 Sep. 70. Rocketdyne, Canoga Park, California. |
| 1971 | Anderson, E. E. | U.S. Patent No. 3,586,515. |
| 1971 | Lockwood, R. M. | U.S. Patent No. 3,592,395. |
| 1971 | Briffa, F. E. J. | U.S. Patent No. 3,606,867. |
| 1971 | Muller, J. L. | "Theoretical and Practical Aspects of the Application of Resonant Combustion |

|  |  | Chambers in Gas Turbines," *Journal of Mechanical Engineering Science*, Vol. 13, No. 3, June, pp. 137–150. |
|---|---|---|
| 1971 | Brown, D. J. | "Noise Emission and Acoustic Efficiency in Pulsating Combustion," *Combustion Science and Technology*, Vol. 3, pp. 51–52. |
| 1971 | Brown, D. J. (Editor) | *Proceedings, 1st International Symposium on Pulsating Combustion*, University of Sheffield, England, Organized by Dept. of Fuel Technology and Chemical Engineering. |
| 1971 | Lockwood, R. M. | U.S. Patent No. 3,618,655. |
| 1971 | Briffa, F. E. J. et al. | A Study of Unvalved Pulse Combustors. Communication 860 presented at the 37th Autumn Research Meeting, Institution of Gas Engineers, London, (Nov.), (Also 2nd Conference on Natural Gas Research and Technology, A.G.A., June 1972). |
| 1972 | Zhuber-Okrog, G. | "Uber Die Vorgange in Strahlrohen Mit Pulsierender Verbzennung," Paper presented at the Institute for Thermal Turbomachines, Graz, Austria (see also *Fortschr.-Ber. VD1.2*, Vol. 6, No. 47, 1976). |
| 1972 | Eick, W. K. | "Pulse Jet Engines as a Source of Energy for Auxiliary Power Units—Pulse Gas Turbine Without Compressor," Paper presented at 39th Meeting of AGARD. (Population and Energetics), Colorado Springs, Col. |
| 1973 | Belter, J. W. | U.S. Patent No. 3,738,290. |
| 1973 | Herse, G. | "Ein Beitnag Zur Weiterentwicklung von Pulsationstrie-Biwerken," *Zeitschrift für Flugwissenschaften*, Vol. 21, No. 6, pp. 189–195. |
| 1974 | Handa, N. | U.S. Patent No. 3,792,581. |
| 1974 | Pearson, R. D. | U.S. Patent No. 3,819,318. |
| 1974 | Marzouk, E. S. and Kentfield, J. A. C. | "Pressure-Gain Combustion, A Means of Improving the Efficiency of Thermal Plant," *Proceedings*, 9th Intersociety Energy Conversion Engineering Conference, pp. 1125–1131. Published by ASME. |
| 1974 | Marzouk, E. S. | "A Theoretical and Experimental Investigation of Pulsed Pressure-Gain Combustion," Ph.D. dissertation, Dept. of Mech. Engineering, Univ. of Calgary, Calgary, Alberta, Canada. |

| 1974 | Peters, R. and Borman, G. | "Low Nitric Oxide Emissions via Unsteady Combustion," *Combustion and Flame*, Vol. 22, pp. 259–261. |
|---|---|---|
| 1974 | O'Brien, J. G. | "The Pulsejet Engine: A Review of Its Potential," MS. dissertation, U.S. Naval Postgraduate School, Monterey, CA. |
| 1974 | Kentfield, J. A. C. et al. | "An Air-Breathing Pressure-Gain Combustor Without Moving Parts," Paper 137, 15th International Symposium on Combustion (only extended abstract published). |
| 1974 | Swithenbank, J. et al. | "Some Implications of the Use of Pulsating Combustion for Power Generation Using Gas Turbines," *Journal of the Institute of Fuel*, pp. 181–189. |
| 1974 | Kentfield, J. A. C. | Results of Preliminary Tests of an Air-Breathing Pressure-Gain Combustor. Technical Note, Dept. of Mech. Engineering, Univ. of Calgary, Calgary, Alberta, Canada. |
| 1974 | Hanby, V. I. and Brown, D. J. | "A Residual Fuel Oil-Fired Pulsating Combustor," *Journal of The Institute of Fuel*, pp. 49–51. |
| 1976 | Crowe, R. K. | "Pulsating Combustion Device Miniaturization," MS. dissertation, U.S. Naval Postgraduate School, Monterey, CA. |
| 1976 | Rehman, M. | "A Study of a Multiple-Inlet Valveless Pulsed Combustor," Ph.D. dissertation, Dept. of Mech. Engineering, Univ. of Calgary, Calgary, Alberta, Canada. |
| 1976 | Clarke, P. H. and Craigen, J. G. | "Mathematical Model of a Pulsating Combustor," Paper C54/76, Sixth Thermodynamics and Fluid Mechanics Convention, Institution of Mechanical Engineers. |
| 1976 | Hollowell, G. T. | "Pulse Combustion—An Efficient, Forced Air, Space Heating System," Proceedings of the Purdue Heating, Ventilating and Air Conditioning Equipment Efficiency Conference, pp. 496–503. |
| 1977 | Kentfield, J. A. C. et al. | "A Simple Pressure-Gain Combustor for Gas Turbines," ASME *Journal of Engineering for Power*, Vol. 99, No. 2, pp. 153–158. |
| 1977 | Severyanin, V. S. and Dereshcuk, E. M. | "Future Prospects for the Use of Pulsating Combustion," *Energetika*, Vol. 5, pp. 138–141. |
| 1977 | Kentfield, J. A. C. | U.S. Patent No. 4,033,120. |

1977    Kentfield, J. A. C.    "A New Light-Weight Warm-Air Blower for Rapidly Preheating Cold-Soaked Equipment," ASME Paper No. 77-WA/HT–20.

1978    Ahrens, F. W. et al.    "An Analysis of the Pulse Combustor Burner," *ASHRAE Transactions*, Vol. 84, Part 1, pp. 488–507.

1978    Vogt, S. et al.    Performance Characteristics of a Pulse Combustion Water Heater. Report of the Ray W. Herrick Laboratories of Purdue University, Lafayette, Indiana. Report No. 1, Research Contract No. PRF 0080-54-1288.

1978    Hirose, Y. et al.    Application of Pulsating Combustion to a Radiant Tube, 5th Members Conference International Flame Research Foundation.

1978    Gill, B. S. and Bhaduri, D.    "Prediction of Pressure and Frequency in Helmholtz Pulsating Combustion Systems," *Indian Journal of Technology*, Vol. 16, pp. 171–176.

1978    Reader, G. T.    "Aspects of Pulsating Combustion," *Proceedings*, 13th Intersociety Energy Conversion Engineering Conference, pp. 548–557.

1978    Rehman, A.    "A Theoretical and Experimental Study of the Affect of Size on Liquid-Fueled, Carburetted Valveless Pulsed Combustors," M.Sc. dissertation, Dept. of Mech. Engineering, Univ. of Calgary, Calgary, Alberta, Canada.

1978    Giammar, R. D. and Putnam, A. A.    "Noise Reduction Using Paired Pulse Combustors," AIAA *Journal of Energy*, Vol. 2, No. 5, pp. 319–320.

1979    Kentfield, J. A. C. and Read, M. A.    "A Pressure-Gain Combustor Utilizing Twin Valveless Pulsed-Combustors Operating in Antiphase," Paper 799378, *Proceedings*, 14th Intersociety Energy Conversion Engineering Conference, pp. 1774–1779. Published by American Chemical Society.

1979    Cronje, J. S.    "An Experimental and Theoretical Study, Including Frictional and Heat Transfer Effects, of Pulsed, Pressure-Gain Combustion," Ph.D. dissertation, Dept. of Mech. Engineering, Univ. of Calgary, Calgary, Alberta, Canada.

1979    Hollowell, G. T.    U.S. Patent No. 4,164,210.

1979    Smay, E. V.    "High Efficiency Home Heating, Part II: Gas Fired Systems," *Popular Science*, Nov., p. 61.

| 1980 | Vogt, S. T. et al. | "Performance of a Pulse Combustion Gas-Fired Water Heater," *ASHRAE Transactions*, Vol. 86, Part 1. |
|------|-------------------|------|
| 1980 | Kentfield, J. A. C. et al. | "Performance of Pressure-Gain Combustor Without Moving Parts," *AIAA Journal of Energy*, Vol. 4, No. 2, pp. 56–63. |
| 1980 | Hollowell, G. T. | Commercialization of Pulse Combustion Furnace With Ultra-High Efficiency. A.G.A. Laboratories Annual Report for 1979 to the Gas Research Institute. GRI-79/0029 (Jan. 1980) |
| 1980 | Clinch, J. M. (Editor) | *Proceedings, Symposium on Pulse Combustion Technology for Heating Applications*, Argonne National Laboratories, Nov. 1979. Argonne National Laboratory, Report No. ANL/EES-TM-87 of May 1, 1980 (i.e., 2nd International Symposium on Pulsating Combustion). |
| 1980 | Duffy, G. | "Pulse-Combustion Furnace Promises Up to 45 Percent Efficiency Boost," *Air Conditioning, Heating, and Refrigeration News*, Sept. 15, p. 25. |
| 1980 | Smith, D. A. et al. | U.S. Patent No. 4,221,174. |
| 1980 | Ferguson, F. A. | U.S. Patent No. 4,226,688. |
| 1980 | Ferguson, F. A. | U.S. Patent No. 4,226,670. |
| 1980 | Kitchen, J. A. | U.S. Patent No. 4,241,720. |
| 1980 | Kitchen, J. A. | U.S. Patent No. 4,241,723. |
| 1981 | Huber, L. | U.S. Patent No. 4,259,928. |
| 1981 | Huber, L. | U.S. Patent No. 4,260,361. |
| 1981 | Piterskikh, G. P. et al. | U.S. Patent No. 4,265,617. |
| 1981 | Chiu, H. H. et al. | "Pulse Combustor Noise: Problems and Solutions," *Proceedings*, International Gas Research Conference, Los Angeles. |
| 1981 | Kodolgy, R. J. and Liljenberg, G. W. | Pulse Combustion Residential Water Heater. A.G.A. Annual Report (Nov. 1980–Oct. 1981), GRI Contract No. 5080-345-0369. |
| 1981 | Kentfield, J. A. C. | "Valveless Pulsejets and Allied Devices for Low Thrust, Subsonic, Propulsion Applications," *Proceedings*, Paper 10, AGARD Conference 307, Ramjets and Ramrockets for Military Applications (AGARD-CP-307). |
| 1981 | Goldman, Y. and Timnat, Y. M. | A Critical Review of Pulsating Combustion. Dept. Aeronautical Engineering, Technion, Haifa, Israel, TAE Report No. 468. |
| 1982 | Putnam, A. A. et al. | U.S. Patent No. 4,314,444. |

| 1982 | Belles, F. E. and Griffiths, J. C. | Commercialization of a Pulse Combustion Furnace with Ultra-High Efficiency. A.G.A. Laboratories Annual Report for 1980 to the G.R.I. GRI-80/0131, PB92-24309, Feb. 1982. |
|------|------|------|
| 1982 | Burnham, F. | "Pulse Jet Gains Headway as Dehydration Device,," *Render, pp.* 16, 17, 52, 62. |
| 1982 | Belles, F. E. and Griffiths, J. C. | Pulse Combustion Furnace. Phase II— Advancement of Developmental Technology, Annual Report (Jan.–Dec. 81) to GRI, Contract No. 5014-341-0112. |
| 1982 | Blomquist, C. A. | Experimental Gas-Fired Pulse Combustion Studies. Argonne National Laboratories Report ANL/EES-TM-214. |
| 1982 | Kitchen, J. A. | U.S. Patent No. 4,336,791. |
| 1982 | Ponizy, G. and Wojcicki, S. | "An Experimental and Numerical Study on Re-ignition and Burning Processes in Pulse Combustors," The Combustion Institute, Western States Section, Paper No. 82-71. |
| 1982 | Sran, B. S. | "The Performance of Two Gas-Dynamically Coupled Pulsed-Combustors Operating in Antiphase," M.Sc. dissertation, Dept. of Mech. Engineering, Univ. of Calgary, Calgary, Alberta, Canada. |
| 1982 | —— | *Proceedings, Vol. 1: Symposium on Pulse Combustion Applications*, Atlanta, Georgia, Mar. 1982. GRI-82/0009.2, NTIS PB82-240060 (i.e., *3rd International Symposium on Pulsating Combustion*). |
| 1983 | Toyonago, H. et al. | "Development of New Combustion Technology for High Performance Gas Appliances," International Gas Research Conference, London, IGRCID-12-83. |
| 1983 | Corliss, J. M. et al. | "Aerovalved Pulse-Combustion Systems for High Efficiency Commercial/Industrial Boilers," GRI Report GRI-80/0182, PB84-125061. |
| 1983 | Davi, M. A. et al. | U.S. Patent No. 4,409,787. |
| 1983 | Putnam, A. A. | U.S. Patent No. 4,417,868. |
| 1984 | Hisaoka, S. et al. | U.S. Patent No. 4,472,132. |
| 1984 | Tikhonovich, D. E. et al. | U.S. Patent No. 4,473,348. |
| 1984 | Hemphill, R. J. | *Gas Research Institute Digest* (GRID), Vol. 7, pp. 1, 4–11, May–June. |
| 1984 | Corliss, J. M. et al. | "NO$_x$ Emissions From Several Pulse Combustors," ASME Paper 84-JPGC-APC-2. (Winter Annual Meeting, New Orleans, La.). |

| 1984 | Keller, J. O. et al. | An Experimental Investigation of a Pulse Combustor-Flow Visualization by Schlieren Photography. Sandia Report No. 84-828 (Nov.). |
| 1984 | Brenchley, D. L. and Bomelburg, H. J. | Pulse Combustion—an Assessment of Opportunities for Increased Efficiency. U.S., DOE, Document PNL-5301/UC-95. |
| 1984 | Olorunmaiye, J. A. | Computer Performance for the Numerical Simulation of Highly-Loaded Valveless Pulsed Combustors. Report No. 314, Dept. of Mech. Engineering, Univ. of Calgary, Calgary, Alberta, Canada. |
| 1984 | Zinn, B. T. | "State of the Art Research Needs of Pulsating Combustion," ASME Paper 84-WA/NCA-19. (Winter Annual Meeting, New Orleans, La.). |
| 1984 | Putnam, A. A. | "Historical Overview of Pulse Combustion with Specific Comments Relative to Noise," ASME Paper 84-WA/NCA-20. (Winter Annual Meeting, New Orleans, La.). |
| 1985 | Olorunmaiye, J. A. | "Numerical Simulation and Experimental Studies of Highly-Loaded Valveless Pulsed Combustors," Ph.D. dissertation, Dept. of Mech. Engineering, Univ. of Calgary, Calgary, Alberta, Canada. |
| 1985 | Keller, J. O. and Saito, K. | An Experimental Investigation of an Unsteady Combusting Flow in a Pulse Combustor. Sandia Report No. 84-8022 (Jan. 85) (see also AIAA Paper No. 85-0322). |
| 1985 | —— | "Lasers Help Researchers Determine the Mechanics of Pulsed Combustion," *Res. Dev.*, Apr., pp. 47–48. |
| 1985 | Corliss, J. M. and Putnam, A. A. | Basic Research on Pulse Combustion Phenomena. Annual Report, Nov. 83–Nov. 84, GRI 85/0029, Apr. 1985. |
| 1985 | Westbrook, C. K. | Successive Reignition of Fuel-Air Mixtures and Pulse Combustion. The Combustion Institute, Western States Section, Oct. |
| 1985 | Putnam, A. A. and Merryman, L. | "Pulse Combustor $NO_x$ As Affected by Fuel-Bound Nitrogen," *J. Air Pollut. Control Ass.*, Vol. 35, pp. 1974–1075. |
| 1985 | Zinn, B. T. | "Pulsating Combustion," *Mechanical Engineering*, ASME, Vol. 107, No. 8, Aug., pp. 36–41. |

1985    Keller, J. O.            "The Dynamics of Injection, Mixing and
                                 Combustion in a Valved Pulse Combustor,"
                                 Paper No. WSS/CI 85-38, The Combustion
                                 Institute, Western States Section, Fall Meeting,
                                 Oct. (see also Sandia Report SAND 86-8775).

1985    Reuter, D. et al.        "Periodic Mixing and Combustion Processes
                                 in Gas Fired Pulsating Combustors," The
                                 Combustion Institute, Central and Western
                                 States Joint Technical Meeting, San Antionio,
                                 Texas.

1985    Kentfield, J. A. C. and  "Pulsating Combustion Applied to a Small Gas
        Yerneni, P.              Turbine," ASME Paper 85-GT-52, Mar. (see
                                 also *International Journal of Turbo and Jet
                                 Engines*, Vol. 4, No. 1–2, 1987, pp. 45–53).

1986    Keller, J. O. and        "Measurements of the Combustor Flow in a
        Saito, K.                Pulse Combustor," Sandia Report SAND
                                 86-8758 (see also *Combustion Science and
                                 Technology*, Vol. 53, No. 2–3, 1987, p. 137).

1986    Putnam, A. A. et al.     "Pulse Combustion," *Progress in Energy and
                                 Combustion Science*, Vol. 12, No. 1.

1986    Dec, J. E. and           "The Effect of Fuel Burn Rate on Pulse
        Keller, J. O.            Combustor Tail Pipe Velocities," Presented at
                                 the International Gas Research Conference,
                                 Toronto, Canada, Sept. (see also Sandia Report
                                 SAND 86-8757).

1986    Keller, J. O. and        "Response of a Pulse Combustor to Changes
        Westbrook, C. K.         in Fuel Composition," *Proceedings*, 21st
                                 International Symposium on Combustion,
                                 Munich (see also Sandia Report SAND
                                 86-8631).

1986    Tsujimoto, Y. and        "Numerical Analysis of a Pulse Combustor
        Machii, N.               Burner," *Proceedings*, 21st International
                                 Symposium on Combustion, Munich.

1986    Vishwanath, P. S. and    "Combustion Time Delays in Pulse
        Priem, R. J.             Combustion Burners," Poster Presentation, 21st
                                 International Symposium on Combustion,
                                 Munich.

1987    Barr, R. K. et al.       "A One-Dimensional Model of a Pulse
                                 Combustor," Paper Presented at 2nd
                                 ASME/JSME Thermal Engineering Joint
                                 Conference, Honolulu, HI. (March).

1987    Keller, J. O. et al.     Pulse Combustion: The Importance of
                                 Characteristic Times. Sandia Report SAND
                                 87-8783.

| 1987 | Kentfield, J. A. C. and O'Blenes, M. | "Methods for Achieving a Combustion-Driven Pressure-Gain in Gas Turbines," ASME Paper 87-GT-126. Also ASME *Journal of Engineering for Gas Turbines and Power*, Vol. 110, Oct. 1988, pp. 704–711. |
| 1987 | Bramlette, T. T. | The Role of Fluid Dynamic Mixing in Pulse Combustors. Sandia Report SAND 86-8622. |
| 1987 | O'Blenes, M. J. | "A Small Gas Turbine Equipped with a Second Generation Pulse Combustor," M.Sc. dissertation, Dept. of Mechanical Engineering, Univ. of Calgary, Calgary, Alberta, Canada. |
| 1987 | Kentfield, J. A. C. and O'Blenes, M. J. | "The Application of a Second Generation Pulse, Pressure-Gain, Combustor to a Small Gas Turbine," AIAA Paper AIAA-87-2156. (See also AIAA *Journal of Propulsion*, Vol. 6, No. 2, Mar.–Apr. 1990, pp. 214–220.) |
| 1988 | Kentfield, J. A. C. | "The Feasibility, from an Installational Viewpoint, of Gas-Turbine Pressure-Gain Combustors," ASME Paper 88-GT-181. |
| 1989 | Kentfield, J. A. C. and Fernandes, L. C. V. | "Improvements to the Performance of a Prototype Pulse, Pressure Gain, Gas-Turbine Combustor," ASME Paper 89-GT-277. (See also ASME *Journal of Engineering for Gas Turbines and Power*, Vol. 112, Jan. 1990, pp. 67–72.) |
| 1989 | Speirs, B. C. | "An Experimental Optimisation of a Large, Aerovalved, Multiple Inlet Pulse Combustor," M.Sc. dissertation, Dept. of Mechanical Engineering, Univ. of Calgary, Calgary, Alberta, Canada. |
| 1989 | Olorunmaiye, J. A. and Kentfield, J. A. C. | "Numerical Simulation of Valveless Pulsed Combustors," *Acta Astronautica*, Vol. 19, No. 8, pp. 669–679. |
| 1990 | Fernandes, L. C. V. | "The Secondary Duct Flowfield of a Pulse, Pressure Gain, Combustor," Ph.D. dissertation, Dept. of Mechanical Engineering, Univ. of Calgary, Calgary, Alberta, Canada. |
| 1990 | Kentfield, J. A. C. and Fernandes, L. C. V. | "Further Development of an Improved Pulse, Pressure Gain, Gas-Turbine Combustor," ASME Paper No. 90-GT-84. |

# 9

## INDUCTION AND EXHAUST
## SYSTEM TUNING

Pressure-wave events in the induction and exhaust systems of internal-combustion engines operating on what are basically either Otto or Diesel cycles have one feature in common with wave events in dynamic pressure-exchangers and pulse combustors, namely, the wave events are cyclic in nature. The majority of internal-combustion engines employing either the Otto or the Diesel cycle are of the reciprocating type; exceptions are engines of the Wankel, rotary, kind. The latter are, to date at least, exclusively of the spark-ignition variety employing the Otto, constant-volume combustion cycle. Functionally, in so far as the operation of their induction and exhaust systems is concerned, Wankel engines are, to a first approximation, indistinguishable from conventional reciprocating, four-stroke, spark-ignition, internal-combustion engines.

For the case of simple dynamic pressure-exchangers the mechanical design of such devices is arranged specifically to maximize the effectiveness of the appropriate pressure-wave cycle being implemented. Possible exceptions are pressure-exchangers that combine intimately, within the one rotor, the additional functions of a turbocompressor and/or a power turbine. The design of pulse combustors must take into account not only the requirements of the gas dynamics but also those of the essential combustion process. The mechanical design involves, therefore, a compromise between the twin needs of the gas dynamics and combustion aspects of such systems. In the case of reciprocating, and Wankel type, internal-combustion engines the wave events are, even when attempts are made to optimize these to yield improved engine performance, subordinate to the main aspects of engine design. Reasons for this include the dominant need to contain, safely, the very high combustion pressures generated and, in most cases, the requirement that the output power be transmitted mechanically, usually by means of a crankshaft, to the load being driven by the engine, machinery dynamics, etc.

Whether or not an attempt is made to utilize, to advantage, wave events in the induction and exhaust systems of reciprocating and Wankel type internal-combustion engines, wave events occur due to the intermittent nature of the flows into, and from, such engines. Similar comments can be made, incidentally, with respect to reciprocating and rotary positive-displacement gas compressors. For the case of internal-combustion engines, wave events can affect engine

236

operation and performance either adversely or positively. An example of an adverse influence is the upsetting of carburetion on carburetted spark-ignition engines; another results in a reduction of engine volumetric efficiency. Wave action can be minimized by employing induction and exhaust passages of relatively large cross-sectional area and, in the case of exhaust systems particularly, capacities, or volumes, serving as decouplers, rather in the manner indicated for a pulse combustor in Fig. 8.15, close to engine exhaust ports. In systems specifically designed to employ pressure-wave events advantageously, induction and exhaust systems having relatively small cross-sectional areas are usual. The lengths of both inlet and exhaust tracts, and in some cases also the cross-sectional area schedule, or variation, as a function station, are carefully optimized.

The analysis of nonsteady flows in reciprocating, and Wankel, engine induction and exhaust systems does not require the introduction of any inherently new theoretical considerations not already covered in the foregoing chapters. Indeed much of the discussion concerning the basic operation of tuned inlet and exhaust systems can be conducted employing only elementary analytical concepts such as those introduced in Chapter 2. Naturally, for fully comprehensive, accurate, performance predictions use has to be made of the complete method-of-characteristics or some equivalent procedure. Comparisons are presented at the end of this Chapter between results obtained experimentally from model exhaust systems with corresponding predictions using the method-of-characteristics.

### 9.1 Reasons for Using Tuned Systems

The reasons for using tuned induction and exhaust systems depends, to some extent, upon the type of engine under consideration. The intended purpose of a tuned inlet/exhaust system on a four-stroke engine differs from that of a tuned exhaust system applied to a two-stroke engine, particularly a very simple two-stroke engine employing piston-controlled porting.

In an application to a naturally aspirated single-cylinder four-stroke engine the wave events in the exhaust pipe are intended to:

a. reduce the exhaust back pressure on the engine piston as the piston moves, in the cylinder, from the bottom (or inner) dead-center (BDC) position toward top (or outer) dead-center (TDC) during the exhaust stroke,

b. maintain a low pressure in the engine exhaust port to encourage removal of exhaust products from the engine combustion, or clearance, space during the short period, close to the piston TDC position on the exhaust stroke, when both the inlet and exhaust valves are partially open (valve overlap).

The intended, and relatively important, function of the wave events occurring

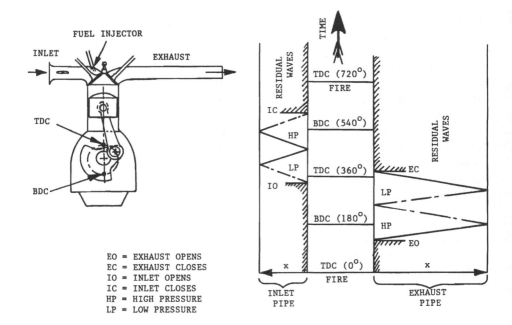

**Fig. 9–1** A simple, single cylinder, four-stroke engine with tuned induction and exhaust pipe (diagrammatic).

in the induction pipe is to "supercharge" the engine by utilizing the inertia of the airflow in the induction pipe entering the engine cylinder. The supercharging action occurs during the latter part of induction as the piston approaches, and passes through, the BDC position. The inertia-generated overpressure is normally arranged to remain imposed until the inlet valve is fully closed.

Figure 9.1 depicts, diagammatically, a simple, single cylinder, four-stroke engine with tuned inlet and exhaust pipes. No obstructions, for example an intake air filter or an exhaust muffler, or silencer, are allowed for. The engine is, therefore, characteristic of a single cylinder, fuel injected, racing engine running at full throttle. The cylinder events, such as the TDC and BDC positions and valve openings and closings, are marked, in chronological order, on an upward-running time scale, the zero of which corresponds to the piston TDC position on the compression stroke. This form of presentation allows the wave events in the exhaust and inlet pipes to be depicted on their respective $t$-$x$ planes to the right and left, respectively, of the column depicting the sequence of events in the engine cylinder. The pressure waves are represented only in a simplistic, basic, form. A single solid line represents a compression wave and an expansion wave is approximated as a single chain-dotted line. It can be seen that in the most elementary sense four basic wave events, each featuring two compression and two expansion waves, occur in both the inlet and the exhaust pipes.

It is clear from the wave-diagram portion of Fig. 9.1 that for a given engine speed, and hence a prescribed cycle duration, both the inlet and the exhaust pipe lengths have to be selected appropriately to generate correctly sequenced periods of high and low pressures at the engine inlet and exhaust valves, respectively. Hence the full benefit of tuned inlet and exhaust systems can only be realized at one particular engine speed and that the beneficial influences of such tuning are generally only realizable over a comparatively small range of engine speeds. This situation is usually acceptable for racing engines where engine speed can be constrained within a relatively narrow band by, for example, frequent gear changing in road-vehicle applications. Inlet and exhaust system tuning can be applied more generally than to racing engines for engines required to operate only, or mostly, at one speed in, say, some marine applications or for driving constant speed alternators.

Two-stroke engines, particularly those of simple design in which both the exhaust and transfer functions are controlled by means of piston-regulated porting in the cylinder walls, are especially sensitive to tuning of the exhaust system. Here the benefit derived is due not only to a lowered exhaust port pressure during the mid-to-latter stages of exhaust but also, most importantly, due to a final increase of pressure in the exhaust port. The latter is due to a compression wave propagating in the exhaust pipe arriving at the exhaust port shortly before it is closed by the piston as the latter moves toward the TDC position. The significance of this pressure increase is that it inhibits a loss of unreacted air/fuel mixture through the open exhaust port. In some cases the pressure increase is sufficient to cause a flow reversal with the consequence that unreacted mixture that had earlier left the engine cylinder via the exhaust port is pushed back into the engine cylinder. A simple two-stroke engine with piston-controlled ports provided with a tuned exhaust system is illustrated diagrammatically in Fig. 9.2. In practice, the transfer passage would not be directed toward the exhaust port as implied in Fig. 9.2.

Pressure-wave events in the exhaust pipes of engines can be improved by arranging for the last section of the exhaust to be divergent as illustrated in Fig. 9.3 for the four-stroke engine considered previously. The divergent feature serves partly as a diffuser and allows a lower pressure to be achieved at the engine exhaust than would otherwise be the case. Divergent tailpipes are often used, also, on highly loaded pulse combustors; examples are shown in Figs. 8.6 and 8.10.

The application of a divergent tailpipe to a two-stroke engine is depicted in Fig. 9.4(a). Usually the arrangement is extended by the addition of a reversed cone culminating in a comparatively small-bore final exhaust pipe; the latter is usually termed a "stinger." This arrangement, shown in Fig. 9.4(b), retains some of the advantages of the divergent exhaust while permitting an exhaust outflow that is closer to a steady flow, and hence less noisy, than the very transitory flow that would otherwise emerge from the open end of a divergent pipe as

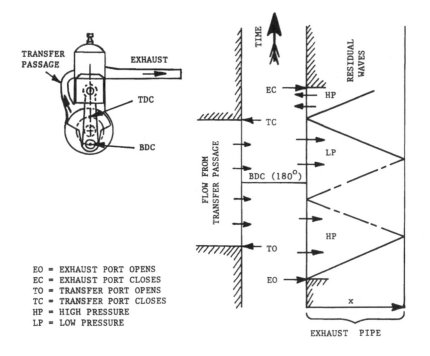

EO = EXHAUST PORT OPENS
EC = EXHAUST PORT CLOSES
TO = TRANSFER PORT OPENS
TC = TRANSFER PORT CLOSES
HP = HIGH PRESSURE
LP = LOW PRESSURE

**Fig. 9–2** A simple, single cylinder, two-stroke engine with a tuned uniform cross-sectional area, exhaust pipe (diagrammatic).

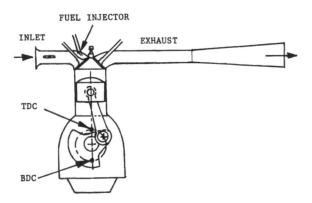

**Fig. 9–3** A simple, single cylinder, four-stroke engine with a tuned induction passage and a tuned exhaust with a divergent, diffuser type, exit (diagrammatic).

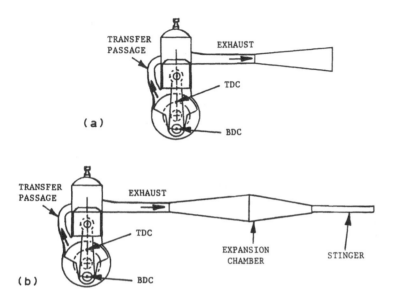

**Fig. 9–4** Simple, single cylinder, two-stroke engines with tuned exhaust systems (diagrammatic).

illustrated in Fig. 9.4(a). The configuration of Fig. 9.4(b) can also be arranged to generate a stronger exhaust-port pressure rise, by virtue of the exhaust-driven pressurization of the enlarged volume, or expansion chamber, that can be obtained using the arrangement of Fig. 9.4(a). A detailed study, by Karnopp et al. (1975), is available of the performance of a system of the type shown in Fig. 9.4(b).

## 9.2 Approximate, Quantitative, Evaluation

While approximate procedures are not usually recommended in preference to method-of-characteristics based analyses, nevertheless very simple evaluations can be used to obtain at least an indication of the lengths, and diameters, of very elementary tuned systems such as those illustrated in Figs. 9.1 and 9.2.

The simplest, and least accurate, of approximate evaluations of duct lengths is based on the assumptions that gas velocity can be ignored while pressure waves travel at the acoustic velocity of the medium. Applying a concept of this nature to a simple four-stroke racing engines shows that, for example, with a representative exhaust gas temperature of 1200 K and an air inflow temperature of 280 K, and exhaust and inlet valve opening durations each corresponding to 260° of crankshaft rotation, the required exhaust and inlet lengths are, at 9000 rev/min engine speed, approximately 0.8 m (2.6 ft) and 0.4 m (1.3 ft), respectively. Such lengths are usually quite easy to accommodate without presenting unusual engine installation problems.

Pipe cross-sectional areas can be selected on the basis of the assumption that the mean flow Mach number should not exceed 0.4 in either the inlet or the exhaust tract during the periods that these communicate with the engine cylinder. The higher the flow Mach number the stronger the wave action but also the greater are the friction losses, etc. A mean flow Mach number of 0.4 is a "rule of thumb" compromise merely to serve as an initial design guide. A consequence of ignoring gas velocities, wave fanning, and finite valve, or port, opening and closing rates when estimating the inlet and exhaust pipe lengths by the method suggested here is that the length of these pipes is typically overestimated substantially. A flow Mach number of 0.4, when brought to rest transiently, can give rise to an increase of the flow absolute stagnation pressure of about 50 percent or an increase of absolute static pressure of about 70 percent. Thus it is possible to achieve, during the latter stage of an induction process, a significant supercharge effect, although it should be remembered that the pressure increase inside an engine cylinder will tend to be lower due to the throttling action of the partly closed, and closing, inlet valve(s).

Based on equal engine speeds, the tuned length of the exhaust pipe of a simple ported two-stroke engine, such as that of Fig. 9.2, tends to be significantly less than that for a four-stroke engine. Typically the length of a two-stroke exhaust pipe is about 60 percent of that of a four stroke operating at the same speed due, primarily, to the:

a. smaller included angle of between about 180 to 200 crankshaft degrees over which the exhaust port is open compared with the 260° or more of the exhaust-valve open period of a four stroke,

b. 40° or so occupied by the desired pressure rise event occurring at the end of the exhaust process.

### 9.3    Application to Multicylinder Engines

The dominant problem associated with the application of inlet and exhaust system tuning to multicylinder engines relates to interference effects between cylinders upsetting beneficial wave action. It is, however, sometimes possible, by the use of carefully thought-out designs, to actually induce mutually beneficial wave interactions between cylinders the exhaust, or inlet, systems of which are interconnected. Generally, system complications tend to increase when such features as turbocharging are added to engines with tuned inlets and, in particular, tuned exhaust systems.

#### 9.3.1    Naturally aspirated engines

It is, of course, possible to test a multicylinder engine with tuned inlets and exhausts as an array of single-cylinder engines and hence to provide an individual inlet pipe and an individual exhaust pipe for each cylinder. In practice this is sometimes done; in fact it is usual to provide a multicylinder racing engine with a separate inlet pipe for each cylinder. It is, however, possible to obtain some advantage from interconnecting exhaust pipes in certain cases, for example, when a multicylinder engine has minimum exhaust flow overlap between interconnected cylinders. Thus, for example, the exhausts of 4, 6 or 8, or more, cylinder engines with uniform firing intervals are often interconnected in various ways largely governed by the firing sequence. It is usually less easy, and less helpful from the performance viewpoint, to interconnect the exhausts of engines with large exhaust-flow overlaps and nonuniform firing intervals such as, for example, wide-angle "V" twin four-stroke engines employing a crankpin common to the two cylinders.

Possible ways in which the exhausts can be connected of a four-cylinder four-stroke engine, with a conventional single-plane crankshaft, and a firing sequence of 1, 3, 4, 2 (or alternatively 1, 2, 4, 3) are presented in Fig. 9.5. Figure 9.5(a) shows a configuration in which four individual exhausts enter an enlarged, diffuser-like, outlet pipe. The potential advantage relative to four separate systems of the type shown in Fig. 9.3 is that between them the four cylinders maintain a more or less continuous outflow in the outlet pipe (Fig. 9.5(b)). This results in an increase in strength of the expansion wave

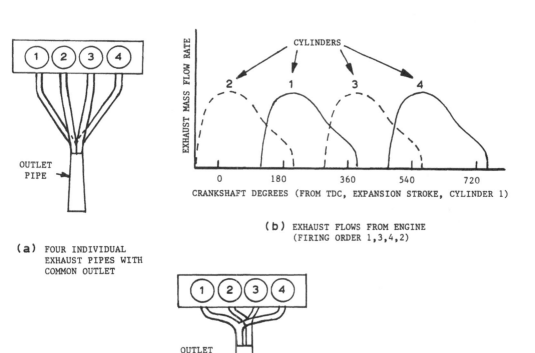

(a) FOUR INDIVIDUAL
    EXHAUST PIPES WITH
    COMMON OUTLET

(b) EXHAUST FLOWS FROM ENGINE
    (FIRING ORDER 1,3,4,2)

(c) EXHAUST PIPES COMBINED INTO TWO
    PAIRS PRIOR TO ENTRY INTO OUTLET PIPE

Fig. 9-5 Tuned exhaust systems for four-cylinder four-stroke engines (diagrammatic).

propagating back toward the exhaust valve during the exhaust phase of each cylinder. A disadvantage, tending to minimize the effectiveness of the arrangement, is the relatively large, 4:1 area ratio, sudden enlargement associated with the junction of the four individual exhausts, only one of which flows vigorously at any instant, and the outlet pipe. The latter problem is diminished in the arrangement shown in Fig. 9.5(c) where the exhausts are connected in two pairs prior to entering the outlet pipe. Because the exhausts are paired of cylinders mutually out of phase by 360° crankshaft degrees there is, as can be seen from Fig. 9.5(b), no exhaust flow overlap and hence the exhaust pipe downstream of each paired connection does not need to be any larger in diameter than the exhaust pipe connected to each cylinder. As a consequence, the nominal area ratio associated with the sudden enlargement at the connector to the outlet pipe is reduced from 4:1 to a much more acceptable 2:1.

### 9.3.2    Engines with mechanical supercharging

Supercharging is only normally applied to multicylinder engines with the possible exception of some single-cylinder research installations. Hence the comments presented here both with respect to mechanically generated supercharge, employing an engine driven blower or compressor, and turbocharging relate to multicylinder engines.

The application of mechanical supercharging to engines with tuned inlets requires the installation of a plenum with which the open ends of the individual inlet pipes communicate. The supercharger outflow serves to maintain a pressure in the plenum, the supercharge pressure, above that of the atmosphere surrounding the engine.

The exhaust system of a mechanically supercharged engine tuned as described previously for normally aspirated engines differs but little from such exhausts. The differences relate to a much stronger initial blowdown, due to higher residual pressures in the cylinder at the point of exhaust-valve opening, and also the elimination of the need to create a low pressure at the exhaust port toward the end of the exhaust process. Scavenge of the engine clearance volume is achieved very effectively, with very small valve overlaps, due to the inlet pressure exceeding that of the exhaust.

### 9.3.3    Turbocharged engines

The application of turbocharging introduces additional constraints when consideration is given to combining inlet and exhaust system tuning with turbocharging or, analogously, supercharging using dynamic pressure exchangers. The presence of the turbine can be modeled, with respect to the influence of the turbine on the exhaust system flow, as if it were a nozzle at the end of the exhaust outlet. Thus, for example, the application of a turbocharger to the exhaust system shown in Fig. 9.5(c) would result in a reduction in the effectiveness of the component labeled "outlet pipe." The area reduction due to the presence of the turbine inlet nozzle would result in the partial reflection upstream, toward the engine exhaust valves, of compression waves propagating

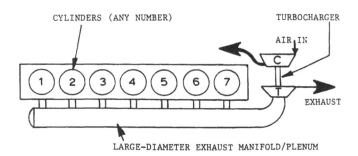

**Fig. 9–6** A multicylinder engine employing constant pressure turbocharging (diagrammatic).

downstream from the engine cylinders. The unwanted upstream reflections could only be avoided if the flow leaving the outlet pipe, and entering the turbine nozzle, was completely steady implying, therefore, either the use of a very large number of cylinders or some form of flow smoothing device such as an exhaust system plenum.

In fact some turbocharged engines, especially those designed to operate at relatively high boost pressure ratios, ignore the influence of pressure waves and employ so-called constant pressure systems. In such configurations each cylinder exhausts into a receiver, or plenum, that also serves as a large diameter manifold interconnecting the engine exhausts with the turbocharger turbine-inlet. An arrangement of this type is shown in Fig. 9.6. Although systems of this type work well under high load conditions a problem can arise, due to turbocharger sluggishness, when attempting to increase, rapidly, the torque of an idling, constant pressure, turbocharged engine. This problem has been considered in more detail by Meier (1971) who describes the reasons for the phenomenon.

Some success has been achieved in combining tuned exhaust systems with turbocharging by dividing the turbine inlet nozzle into sections, each sector being connected to the exhausts from one or more cylinders. Inherent problems with so-called pulse charging systems of this kind are the relatively low turbine efficiency resulting from partial admission, at any instant of time, over only a portion of the total nozzle area of the turbine and the turbine-blade fatigue loading resulting from the partial admission. The former problem has been considered by Meier (1971) and the latter by Kirsten (1971).

It has been shown to be possible, by means of the use of special devices known as pulse converters, to combine tuned exhaust systems with substantially steady, full admission, flow through the turbines of turbochargers. In some cases it is still necessary to divide the turbine inlet nozzle into two or more sectors, however, the flow through each sector approximates, quite closely, steady flow since the exhausts of a number of cylinders, even in some cases with a measure

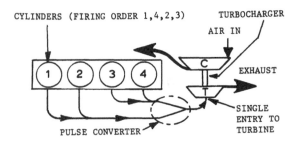

(a)  FOUR CYLINDER TWO-STROKE ENGINE WITH A
      SINGLE PULSE CONVERTER (ZEHNDER, 1968)

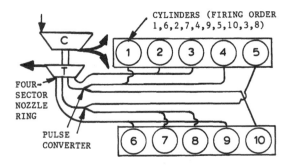

(b)  V.10 FOUR-STROKE ENGINE WITH TWO
      PULSE CONVERTERS (MEIER, 1968)

**Fig. 9–7** Turbocharged Diesel engines using exhaust-system pulse-converters (diagrammatic).

of exhaust pulse overlap, pass through each sector of the turbine inlet nozzle. The most common forms of pulse converter consist of narrow-included-angle Y type exhaust-pipe junctions adapted to serve partly as nonsteady flow ejectors and partly as devices for inhibiting unwanted wave reflections. The arrangements of two pulse-converter equipped exhaust systems, one for an inline four-cylinder two-stroke engine, and the other for a V.10 four stroke, are illustrated in Figs. 9.7(a) and 9.7(b), respectively. The system shown in Fig. 9.7(a) features one pulse converter, that of Fig. 9.7(b) two pulse converters.

### 9.3.4  Pulse converters

The most common form of pulse converter differs from a simple Y type pipe junction, two examples of which appear in Fig. 9.5(c), in that the duct cross-sectional area downstream of the junction is somewhat larger than that of either of the inlet pipes leading to the junction. The downstream duct, known as the mixing tube, typically has a cross-sectional area about 1.5 times that of either of the inlet tubes. Generally, exhaust gases approach the pulse converter simultaneously in both of the inlet passages, however, at any instant one pulsating stream will contain a higher pressure, faster flow than the other. The more energetic of the flows serves, therefore, as the primary flow, the less energetic flow as the secondary stream in the nonsteady ejector like action of the pulse converter. A diffuser can be added, if need be, downstream of the mixing tube to slow the mixed flow and hence increase the static pressure prior to entry into the turbocharger turbine. The roles of driver and driven streams alternate between the two approaching flows due to the generally nonsteady nature of the pulsating flows in each of the pulse-converter inlet ducts.

In order to minimize the chance of backflow, at the pulse converter, from the more energetic to the less energetic stream it is usual to provide a mild contraction, of about 10 to 20 percent of the duct cross-sectional area, in each inflow passage at the point where these amalgamate. This is usually accomplished by provision of a tongue, projecting toward the mixing tube, between the two inlet passages. Figure 9.8 shows a two-inlet pulse converter of a type described by Meier (1968). Descriptions of both steady and nonsteady analogies of the flow fields in pulse converters have been given by Zehnder (1968).

Because the mixing process in a pulse converter is primarily nonsteady in nature it appears to be possible to use very short mixing lengths, without a

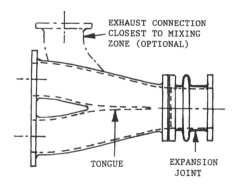

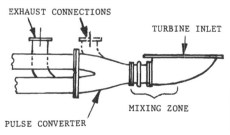

**Fig. 9–8** A two-inlet pulse converter based on a design presented by Meier (Meier, 1968).

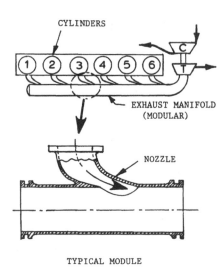

**Fig. 9–9** The exhaust system of a turbocharged Diesel engine equipped with a modular, multiinlet, pulse converter (Magnet and Curtil, 1980).

final diffuser, as indicated in Fig. 9.8. This is particularly useful from the viewpoint of achieving a compact installation. Earlier proposals relating to pulse converters sometimes showed not only a much longer mixing tube but also a downstream diffuser and, additionally, a volume, or plenum, interposed between the diffuser outlet and turbine inlets (Petak, 1964). A theoretical analysis of the operation of pulse converters has been made by Zeller (1965, 1966).

Whilst most ejector-type pulse converters are provided with two inlets, sometimes three, or more, inlets can be used. At least one early pulse converter proposed suggested the use, in one pulse converter, of a large number of inlets (Birmann, 1956). A fairly recent paper, due to Magnet and Curtil (1980), describes a form of modular pulse converter in which each cylinder of a multicylinder engine exhausts into a common manifold via a nozzle. In this type of system the number of inlets is, therefore, equal to the number of cylinders. The fluid in the manifold represents the driven flow that receives nonsteady, jetlike, mass and energy inputs from the exhaust of each cylinder in turn. Figure 9.9 shows, diagrammatically, the modular type pulse converter described by Magnet and Curtil.

## 9.4  Application of the Method-of-Characteristics

Jenny (1950) described the application of the method of characteristics to simulated, model, exhaust systems. He showed, in his classical presentation, predicted results obtained both without, and in some cases with, the inclusion of friction and he also compared his predictions with results obtained experimentally from the model engine-exhaust system. Three configurations were studied all with a simulated valve opening period of 260 crankshaft degrees representative of four-stroke engines. The most elementary arrangement modeled was a simple, straight, exhaust pipe of uniform cross-sectional area, similar to that shown diagrammatically in Fig. 9.1. Another arrangement employed a straight, uniform, duct terminating in an outlet diffuser and was, therefore, similar to the configuration of Fig. 9.3. Jenny also modeled analytically, and tested experimentally, a system in which a uniform, straight, exhaust pipe terminated in a nozzle having an exit area equal to only 32.5 percent of the cross-sectional area of the uniform pipe. The latter configuration was intended to be representative of a single, tuned, exhaust duct connected to the inlet nozzle of the turbine of a turbocharger.

Jenny's results for the simple, straight, uniform pipe, straight pipe plus diffuser, and straight uniform pipe plus nozzle are presented in Figs. 9.10, 9.11, and 9.12, respectively. Dominant features of the diagrams are: the relatively small influence of friction, the very close correlation between experiment and theory and, relative to the simple uniform exhaust pipe, the substantial performance benefit due to the diffuser, and the substantial deterioration resulting from the use of a nozzle-type pipe ending. The latter result serves to emphasize the merit of exhaust systems configured to minimize, unwanted reflections from the turbine, as the pulse-converter concept assists in doing.

The Jenny calculations, the results of which are presented in Figs. 9.10, 9.11, and 9.12, were performed without the use of a digital computer and hence, for brevity, no account was taken of the cyclic nature of the flows in the simulated exhaust systems. For more complicated systems than those studied by Jenny, particularly for engines with many cylinders and interconnected exhaust pipes, it is probably essentialy, certainly very desirable, to employ computerized methods and hence to take into account, very easily, the cyclic nature of the flow system.

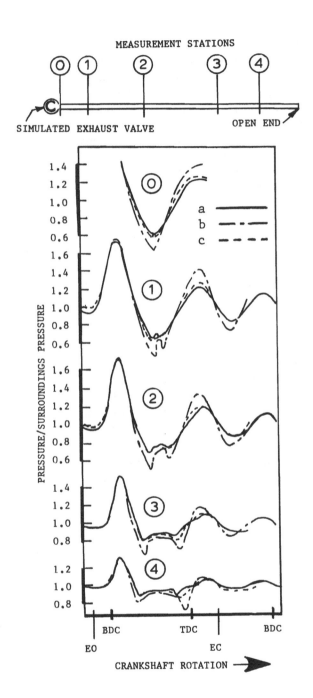

Fig. 9-10 Experimentally measured, and predicted, time dependent pressure variations at several stations in a simulated exhaust pipe (Jenny, 1950). (a) Measured, (b) Predicted (friction ignored), (c) Predicted (influence of friction included).

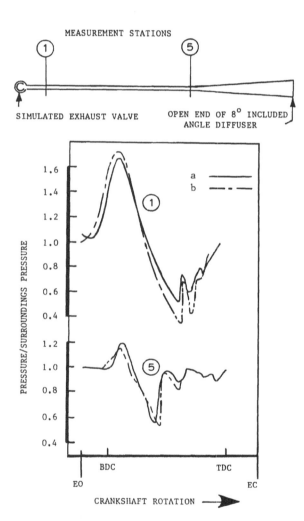

**Fig. 9–11** Experimentally measured, and predicted, time dependent pressure variations in a simulated exhaust pipe with a divergent exit (Jenny, 1950). (a) Measured, (b) Predicted (friction ignored).

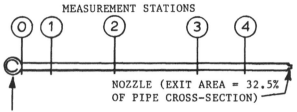

MEASUREMENT STATIONS

NOZZLE (EXIT AREA = 32.5%
OF PIPE CROSS-SECTION)

SIMULATED EXHAUST VALVE

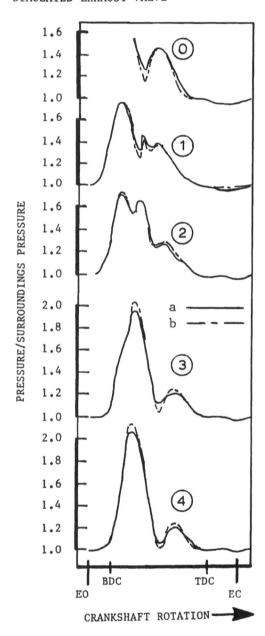

Fig. 9–12 Experimentally measured, and predicted, pressure variations at several stations in a simulated exhaust pipe with a nozzle at the pipe exit to represent a turbocharger turbine (Jenny, 1950). (a) Measured, (b) Predicted (friction ignored).

# 10

## OTHER NONSTEADY
## FLOW DEVICES

Several nonsteady flow devices not described here previously, or not described in significant detail, are available, or have been suggested by various researchers as potentially promising. In most cases a definitive, and adequate, full theoretical treatment permitting a complete understanding of each of the systems considered here does not yet appear to be available. This also seems to be the situation with respect to nonsteady flow ejectors that have been widely used in the blast-pipe systems of steam locomotives. This application, which involves low, and variable, frequency operation of exhaust-steam-driven nonsteady flow ejectors to draw air through locomotive fireboxes dates from the early part of the last century. More recently, nonsteady flow ejectors operating at much higher frequencies have been used in connection with pulse combustors and also reciprocating internal-combustion engine exhaust systems.

### 10.1  Nonsteady Flow Ejectors

So far it appears that nonsteady flow ejectors have only been applied in circumstances in which the available primary flows have been inherently nonsteady. Such applications include the steam locomotive duty mentioned previously and use as ejector type thrust augmenters on pulse jets and other pulse-combustor systems as described, very briefly, in Chapter 8. The so-called pulse converter described in the previous chapter is yet another application in which the primary flow of the ejector-like pulse converter is inherently nonsteady. In the case of the pulse converter, in particular, the secondary flow is also inherently nonsteady. It is, of course, possible, as will be described later, to convert an otherwise steady primary flow into one which, with respect to the operation of any one secondary flow duct of an ejector with multiple secondary flow ducts, is nonsteady. The potential advantage of doing this hinges upon fundamental differences between the operation of steady and nonsteady flow ejectors, the latter being in principle more efficient than the former.

In a steady-flow ejector the process of energy transfer from the primary to the secondary stream depends upon a shearing-action induced flow-mixing process. It is because of this that the secondary flow duct, often of uniform cross-sectional area, is known as the mixing tube, or mixing zone. The process of energy transfer by mixing depends upon both the primary and the secondary fluids being nonideal in that each has a nonzero viscosity. Energy transfer solely by mixing involves significant irreversibilities with, consequently, a low resultant efficiency as demonstrated in numerous analytical models of, and experiments with, steady-flow ejectors. While mixing plays a role in the operation of nonsteady flow ejectors it is subordinate to the main mode of energy transfer between the primary and secondary flows that is by a pressure-exchanger type process. Very crudely the intermittent primary flow can be thought of as forming a succession of "gas pistons" as it passes through the secondary flow duct or, in the terminology of steady-flow ejectors, mixing tube. It appears, from flow visualization studies conducted in connection with the use of nonsteady flow ejectors as thrust augmenters for pulse jets, that the primary fluid mixes, to some extent, with the entrained secondary flow as the plugs of primary fluid slow, and hence expand, to fill the cross section of the secondary duct.

### 10.1.1  Thrust augmentation

Much of the qualitative and quantitative information available on the operation, and performance, of nonsteady flow ejectors was generated during the course of work carried out by the Advanced Research Division or Hiller Aircraft and reported by Lockwood et al. (1960). Lockwood and his associates showed, in part from water-tank flow visualization experiments, that the flow events in a nonsteady flow ejector appeared to follow the sequence presented in Fig. 10.1. Quite independently evidence exists from the observation of the visible exhausts of steam locomotives that flow downstream of a nonsteady flow ejector outlet tends to follow the pattern indicated in Fig. 10.1(a) and (b). It can be seen from Fig. 10.1 that the flow pattern within the ejector is significantly removed from one that can be described adequately as one-dimensional transient flow.

Lockwood et al. (1960) also showed, from experiments using as intermittent subsonic primary flows either the inlet backflow or the exhaust of valveless pulse jets, that an optimum ejector configuration tended toward the geometric arrangement shown in Fig. 10.1. Figure 10.1 shows a bellmouth inlet leading into a divergent section of about 8° included angle having a length-to-throat diameter ratio of approximately 2.5 to 3.0. This corresponds to an area ratio, for the divergent portion of the ejector, of about 1.8 to 2.0. Static-thrust augmentation ratios of up to 2.4 (i.e., 140 percent increase in static thrust) were obtained when the ratio of the throat area of the nonsteady ejector to the exit of the primary jet was 4. By comparison, a steady-flow ejector capable of producing a comparable static-thrust augmentation ratio requires the use of multiple, hypermixing type, primary nozzles in conjunction with an area ratio of approximately 25 instead of the 4:1 ratio of the nonsteady ejector.

Unfortunately it has been found (Marchal and Serventy, 1963; Foa, 1970) that as the velocity of the secondary flow approaching an ejector increases, as, for example, when ejector thrust augmenters are used to augment the thrust of moving vehicles such as aircraft, the thrust augmentation obtainable falls off. Ultimately the ejector is unable to overcome self-drag and hence generates a

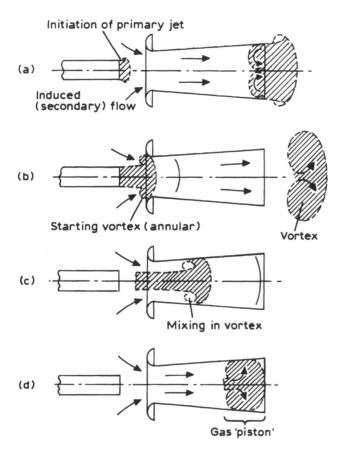

**Fig. 10–1** Apparent sequence of flow events in a pulse jet, nonsteady flow, thrust augmenter, ejector (after Lockwood, Sargent, and Beckett, 1960).

negative net thrust. Despite such problems ejector type thrust augmenters are useful in environments in which the approach flow velocities are low. Examples of the latter type appear in Figs 8.10 and 8.13 where nonsteady flow ejectors are employed as pumping devices rather than as thrust augmenters in the direct jet propulsion sense.

### 10.1.2   Primary-fluid fluidic oscillator

The writer was, at one time, involved with, and was partly responsible for initiating, a project in which an attempt was made to take advantage of the apparent benefits of nonsteady flow ejectors when only a steady flow of fluid is available as a primary source. The system employed a heavy-current fluidic oscillator the output of which served to provide two, alternating, intermittent,

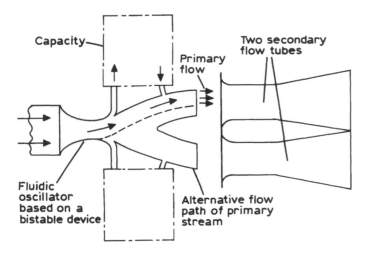

**Fig. 10–2** Nonsteady-flow ejector employing a fluidic oscillator as a primary-flow distributor (diagrammatic).

primary streams each injecting into an individual mixing tube. The apparatus is shown, diagrammatically, in Fig. 10.2.

Functionally the fluidic oscillator served to accept a uniform, steadily flowing, high-pressure flow that was diverted, alternately, into the two outlets of the bistable device that, by virtue of the feedback circuits shown chain dotted in Fig. 10.2, served as an oscillator. The complete system had the advantage, in common with a conventional steady-flow ejector, of not incorporating any moving parts. The main problem experienced during the very short term, low budget, test program was that irreversibilities in the fluidic oscillator canceled out any gains that should otherwise have been realized from the nonsteady pumping action. There may be a potential for achieving a significantly improved performance and hence, for a given duty a reduction in the primary flow rate required, by developing a superior fluidic oscillator. An alternative, at the price of introducing one moving part, might be to substitute some form of mechanical distributor valve for the fluidic oscillator.

### 10.1.3    Valved primary flow

Figure 10.3(a) shows, diagrammatically, a nonsteady flow ejector provided with a central, rotary, primary-fluid distributor valve supplying, sequentially, a number of primary nozzles each of which communicates with a corresponding, aligned, secondary flow tube. Such a system should be configured in a manner that ensures that regions of strong nonsteady flows are, as far as possible, restricted to the primary nozzles themselves and the flow fields downstream of these.

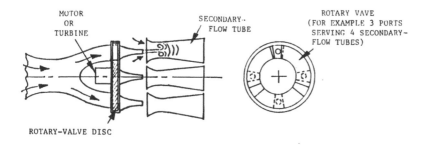

MOTOR
OR
TURBINE

SECONDARY-
FLOW TUBE

ROTARY VAVE
(FOR EXAMPLE 3 PORTS
SERVING 4 SECONDARY-
FLOW TUBES)

ROTARY-VALVE DISC

(a)   ROTARY VALVE

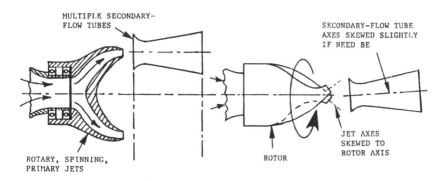

MULTIPLE SECONDARY-
FLOW TUBES

SECONDARY-FLOW TUBE
AXES SKEWED SLIGHTLY
IF NEED BE

ROTARY, SPINNING,
PRIMARY JETS

ROTOR

JET AXES
SKEWED TO
ROTOR AXIS

(b)   SPINNING JETS

**Fig. 10–3** Nonsteady-flow ejectors with valved, or spinning, primary jets.

An alternative approach, in which the flow within the primary nozzles is completely steady relative to the nozzles, involves the use of spinning primary nozzles. An ejector of this type is shown in Fig. 10.3(b). The rotor carrying the primary nozzles, or nozzle, can be made self-driving by skewing the centerlines of the primary jets relative to the centerline of the rotor body. This feature is illustrated in the right-hand portion of Fig. 10.3(b) for a rotor on which are mounted two, diametrically opposite, primary nozzles. If rotor bearing friction were absent, and also provided there were no internal or external windage losses in the rotor, the secondary flow ducts would be arranged with their axes parallel to the rotor centerline. With mechanical and flow frictional forces active on the rotor the axes of the secondary flow tubes should be skewed, in the manner shown in the right portion of Fig. 10.3(b), such that the flow within each secondary tube is parallel to the axis.

An important feature of the type of ejector illustrated in Fig. 10.3(b) is that, due to the motion of the primary jets, the location of a primary jet feeding a secondary tube moves from one side of the secondary tube to the opposite side as the primary jet traverses the secondary tube inlet. This results in a significant skewing of the transient flow field in each secondary tube. This is in distinct contrast to the flow symmetry, in the transverse direction, with other nonsteady flow ejectors, for example that shown in Fig. 10.3(a).

### 10.1.4   Analytical treatment

At least one attempt has been made to apply the numerical, method-of-characteristics-like procedure described in Chapter 5 to the analysis of a nonsteady flow ejector. The configuration analyzed was that of a secondary flow duct of a pressure-gain combustor of the type shown in Fig. 8.13. The secondary flow duct of the pressure-gain combustor featured a nonsteady flow ejector at the upstream end communicating with a long passage leading to the component labeled, in Fig. 8.13, "combining cone." A restrictor plate was located at the exit of the long, secondary flow duct.

Not surprisingly, considering that it was based only on the consideration of a one space-dimensional time-dependent flow, the prediction overestimated performance significantly relative to corresponding test data; however, the trends predicted did agree, to some extent, with those observed experimentally. The parameter varied during the experiments, and also when making performance predictions, was the magnitude of the blockage due to the "restrictor plate" identified in Fig. 8.13. The experimental apparatus employed several interchangeable restrictor plates of different blockages.

Shortcomings of the analysis appear to relate to the treatment of the entering primary and entrained flows and also to the omission of the possibility, particularly with large restrictor-plate blockages, of the transient spillage of primary fluid into the air-inlet plenum. Any fluid spilled in this way must, of course, be subsequently reingested with either the ejector secondary flow, the pulse-combustor inflow, or both. Certainly the indication is further work is required before a satisfactory analysis of the performance of the ejector and attached secondary flow duct extension can be obtained. A more detailed summary of the analytical procedure described here only very briefly is available (Kentfield and Fernandes, 1990(b)).

## 10.2   Crypto-Steady Pressure-Exchange

Crypto-steady pressure-exchange is a concept due to Foa (1955) in which processes involving the principle of pressure-exchange occur in devices that relative to at least one frame of reference appear to be of the steady-flow type. The term "crypto," derived from a Greek word meaning "hidden" or "secret," reflects the sometimes obscure nature of the orientation of the frame, or frames, of reference in which the flow field can be regarded as steady. Usually a crypto-steady pressure-exchanger comprises a freely spinning rotor, similar to that shown in Fig. 10.3(b), from which issue jets, or sheets, of driving fluid. The sheets of driving fluid, termed pseudo-blades, are so arranged that they induce a flow in a fixed duct or channel. The jets issuing from the rotor can be

SOLID ARROWS REPRESENT
ABSOLUTE VELOCITY OF THE
PRIMARY FLOW PRIOR TO
PRESSURE EXCHANGE AND
MIXING  (LOSS-FREE ROTOR)

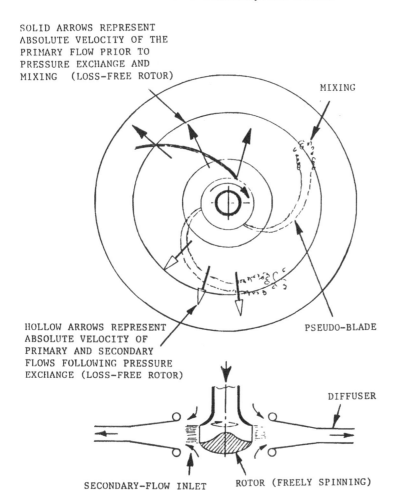

MIXING

HOLLOW ARROWS REPRESENT
ABSOLUTE VELOCITY OF
PRIMARY AND SECONDARY
FLOWS FOLLOWING PRESSURE
EXCHANGE (LOSS-FREE ROTOR)

PSEUDO-BLADE

DIFFUSER

SECONDARY-FLOW INLET        ROTOR (FREELY SPINNING)

**Fig. 10–4** A radial-outflow type crypto-steady pressure-exchanger ejector (diagrammatic).

orientated to induce either a radially outward or an axial flow, or a combination of these, i.e., a conical flow, the fixed, or stationary, duct work being arranged appropriately. Figure 10.4 shows, diagrammatically, a simple, three primary-jet, radial flow crypto-steady pressure-exchanger configured to perform as a primary-to-secondary, or induced flow, pump in the manner of an ejector.

The transfer of energy from the primary to the secondary fluid takes place first by means of a pressure-exchange process in which the pseudo-blades interact with the secondary fluid effecting a momentum exchange between the primary and secondary flows. Mixing will, of course, also occur, to some degree, in parallel with the pressure-exchange process. Ultimately the pseudo-blades will break down and mix with the secondary fluid but, as in a well-designed nonsteady flow ejector, the main mode of energy transfer is by pressure-exchange. To an observer stationary relative to the rotor the flow field appears to be steady whereas to any other observer it is nonsteady. A particular case would be a stationary observer.

For a hypothetical situation in which the rotor runs on frictionless bearings, and also assuming that the freely spinning rotor is without internal or external windage losses, each particle of fluid leaving the rotor will, relative to a stationary frame of reference, be traveling radially outward as implied in Fig. 10.4. Taking into account the concept of the conservation of angular momentum, and the requirement that the primary and secondary streams remain in physical contact after pressure exchange, it can be shown, bearing in mind that the secondary fluid entered the system without acquiring whirl, that the exiting flow leaves the device purely radially.

Substituting an annular duct, as shown in Fig. 10.4, for the discrete, multiple, secondary flow tubes of the apparatus shown in Fig. 10.3(b) converts it into an axial flow version of the radial flow crypto-steady pressure-exchanger ejector depicted in Fig. 10.4. An axial flow ejector is often a more useful configuration, particularly for applications such as thrust augmentation, than one of the radial flow type. Actually the process of conversion would probably also require that the round primary jets, as depicted in Fig. 10.3(b), be arranged more in the form of narrow radial slits to permit the pseudo-blades to fill the full height of the annulus. Such a modification has been implied in Fig. 10.5. The change from the Fig. 10.3(b) configuration to that of Fig. 10.5 also serves to remove probable problems associated with the inclined, or skewed, flow fronts in each of the multiple secondary flow tubes of the ejector of Fig. 10.3(b). Losses that

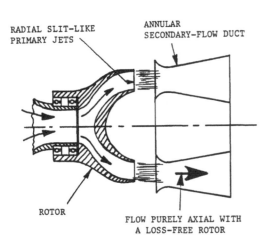

RADIAL SLIT-LIKE PRIMARY JETS

ANNULAR SECONDARY-FLOW DUCT

ROTOR

FLOW PURELY AXIAL WITH A LOSS-FREE ROTOR

Fig. 10–5 An axial-flow type crypto-steady pressure-exchanger ejector (diagrammatic).

could also arise due to periodic interruption of the primary flow as it traverses from one secondary tube to the next are avoided with the apparatus of Fig. 10.5.

### 10.2.1  Thrust augmentation

Most of the available experimental, and also predicted, performance data relating to crypto-steady pressure-exchangers appear to be devoted to thrust augmentation applications. Both tests and performance predictions have been made for situations involving like and also unlike primary and secondary fluids. Examples including air driving air, water driving water, and a steam primary fluid acting on water. The water secondary fluid arrangements relate to marine-propulsion applications. Thrust augmentation ratios obtained experimentally with a water-water system were as high as 2.35, or about 70 percent of the ideal augmentation ratio, for a crypto-steady thrust augmenter having a secondary to primary-flow area ratio of 38:1 (Foa, 1970). With a steam primary flow and a water secondary flow a static thrust augmentation ratio of 4.85 was obtained falling to 4.25 at a condition equivalent to a vessel traveling at 16 knots (8.23 m/s). The foregoing and other data reported by Foa (1970) showed not only that the experimentally obtained performance was, in many cases, close to that predicted theoretically but also the importance, as noted previously in Section 10.1.1, of maintaining the highest possible ratio between the velocity of the primary jet to that of the vehicle.

## 10.3  Detonation-Wave Combustors

Many workers have, over an extended period of approximately fifty years, proposed devices in which detonation waves can be employed for useful purposes. Some of these proposals have been supported by preliminary experimental work and subsequent comparison of the results obtained with detonation wave theory. Much of the established theory relating to detonation waves dates from the last part of the nineteenth century and the early part of the twentieth century.

The theory of detonation waves will not be presented here since it is already available in a number of works. For example, a presentation of the basic theory of detonation waves has been given by Foa (1960) and a more elaborate treatment is that due to Shchelkin and Troshin (1965). In summary, a detonation wave is caused by a very strong shock wave propagating into a combustible mixture, usually, for convenience, assumed to be at rest. Typically the initiating shock wave has a propagation Mach number, referred to the unreacted combustible mixture, in the range from about 3 to 10 depending upon the temperature of the combustible mixture, the calorific value of the fuel, the mixture strength, etc. Combustion follows, very rapidly, the passage of the initial shock that serves to raise the temperature of reactants and hence cause ignition. The static pressure falls, during combustion, from an initial value, $p_2$, determined, as a multiple of the pressure $p_1$ of the unreacted mixture upstream of the shock by means of normal shock relations, to a much lower value. On the basis of the most simple theoretical treatment the final static pressure $p_3$, of the fully reacted products is given by:

$$p_3 - p_1 = \tfrac{1}{2}(p_2 - p_1)$$

The physical reason for the fall in pressure associated with the combustion reaction follows from consideration of the consequences of the increase in temperature due to combustion. On the basis of accounting for the requirements of continuity the higher temperature fluid at state 3 must move more slowly than the unreacted material at state 2. The implementation of a reduction in flow velocity between states 2 and 3 requires that there be a reduction of pressure. Therefore the combustion reaction is coincident with an expansion process. It can also be shown that for a self-sustaining, continuing, detonation wave the complete combustion zone propagates, relative to the fully reacted products of combustion, at a velocity equal to the speed of sound in the products of combustion. This condition is known as the Chapman-Jouguet condition and is a fundamental feature of a propagating, fully self-sustaining detonation wave. Detonation waves of the continuing, self-sustaining type are known as Chapman-Jouguet detonation waves. Transiently, for very short time intervals, detonations can exist that do not meet, in exact detail, the Chapman-Jouguet conditions. Figure 10.6 is an attempt to show three Chapman-Jouguet detonation waves propagating to the right, in a tubular reactor of uniform cross-sectional area, in stationary combustible mixtures of a typical hydrocarbon fuel in air. For case (a) of Fig. 10.6 the reactants are assumed to be a temperature, $T_1$, of 300 K. For case (b) the temperature $T_1$ was taken as 1200 K, a temperature at which the reactants may start to react spontaneously at a significant rate indicating a rather extreme circumstance. Case (c) represents a situation for which $T_1$ is also 1200 K but with a mixture strength of only 70 percent of stoichiometric.

The information of Fig. 10.6 should be taken only as indicative since actual detonation waves are very sensitive to fuel properties, oxygen content, etc. Nevertheless, the diagram serves to illustrate the very high propagation velocities obtained, the very high-pressure ratios achieved with detonations propagating into reactants at room-like temperature [case (a)], the significant influence of reactant temperature [case (b)], and also the influence of inerts on detonation waves [case (c)].

It is worth noting that the final at-rest conditions, state 4 in Fig. 10.6, are those that can be expected from combustion at constant volume, and at rest, of the reactants at state 1. From this viewpoint a detonation wave can be thought of, therefore, as a mechanism for achieving, in effect, a very rapid constant-volume combustion reaction. It may be, however, that the advantage of a high reaction rate is somewhat undermined, in cyclic devices, due to the time required to scavenge the reactor and to replenish it with new reactants.

A detonation wave can either be initiated, following ignition, gradually in a tube filled with a suitable combustible mixture, for example by roughening the tube interior, or it can be started almost instantaneously by means of a

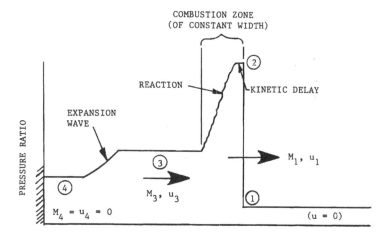

Fig. 10–6 The propagation of detonation waves in a tubular reactor of uniform cross-sectional area. Typical hydrocarbon fuel/air reactants (indicative).

high-energy igniter. A high-energy igniter sometimes consists of a small diameter tube, communicating with the detonation reactor, charged with fuel and oxygen. The fuel-oxygen mixture is ignited by means of an ordinary automotive type spark plug. In the case of a gradual, naturally forming, detonation wave ignition of the reactants in the detonation tube can be direct by means, for example, of a spark. This method has the disadvantage, compared with the use of a high-energy igniter, in that the combustion front may travel, relatively slowly, for a significant distance into the reactor before a detonation wave forms. A more detailed description of oxygen-enriched high-energy igniters, or drivers, has been given by Helman et al. (1986).

The dominant uses proposed for detonation-wave reactors are as gas-turbine combustors or nonresonant type pulse jets. The proposals have ranged from combustors of annular form, in which detonation waves travel around the interior of a toroidal duct, to linear devices in which periodic detonations traverse, either longitudinally or transversely, straight passages.

### 10.3.1  Toroidal-flow combustors

A prototype toroidal, or annular, detonation tube reactor has been described by Edwards (1976). This device, illustrated diagrammatically in Fig. 10.7, has been proposed as, essentially, the basis for a gas-turbine combustor. It has been suggested, by Edwards, that the toroidal detonation tube could constitute the primary zone of a gas-turbine combustor having a very uniform circumferential temperature distribution.

Tests with a toroidal detonation tube showed that either two or sometimes three equispaced detonation waves circulated continuously, depending upon operating conditions, within the passage. The reactor was fed with fuel and oxidant through continuous slots on one side of the toroid while products of combustion were withdrawn from a slot on the opposite side of the toroid. The inlet and exhaust slots are apparent in Fig. 10.7. The valve actions of the inlets and the exhaust were controlled by regulating the fuel and oxidant supply pressures.

The device was found to operate according to the classical detonation wave theory that was used to establish the tabulated evaluation shown in Fig. 10.6. A fundamental difference between the toroidal detonation-tube conditions and the situation presented in Fig. 10.6 was that the reactants instead of being at rest, as indicated in Fig. 10.6, moved toward the advancing detonating waves at about 500 m/s, giving detonation-wave propagating velocities relative to a fixed datum in the region of 1500 m/s instead of values in the region of 2000 m/s, which would be expected on the basis of the gas state of the reactants in the toroid had these been at rest. It may be possible, ultimately, by utilizing the residual energy of the combustor exhaust gases to develop the toroidal detonation wave system into a pressure-gain combustor. It is worth noting that the toroidal combustor is a crypto-steady device (Section 10.2). To an observer

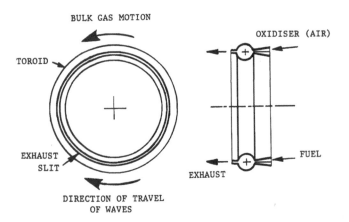

**Fig. 10–7** Annular, or toroidal, detonation-wave combustor (Edwards, 1976).

moving with the detonation waves, the flow field appears to be steady, whereas to, say, a stationary observer, the flow field is nonsteady.

### 10.3.2    Combustors operated cyclically

The majority of devices of this class consist of a straight reactor tube of uniform cross-sectional area, closed at one end, with a valve, or valves, built into the closed end, for admitting, periodically, a new charge of fuel and oxidant. In some systems the flow of oxidant enters through narrow, permanently open, ports adjacent to the closed end or is due to backflow through the open end of the reactor. An ignition system is provided in the vicinity of the closed end of the tube. A generic detonation combustor of the foregoing type is illustrated in Fig. 10.8.

If ignition is by a conventional, relatively low energy spark, the reactor tube is almost inevitably fairly long and is only capable of operating at low frequency. Nicholls et al. (1957) describe a laboratory prototype of a nonresonant, detonative, pulse jet conforming to that description. The Nicholls et al. unit was of nearly 2 m length and had a reactor tube bore of 24 mm. The firing frequency of the Nicholls et al. hydrogen/oxygen pulse jet was in the region of 30 Hz or less. A more recent proposal, due to Edwards et al. (1969) features multiple, parallel reactor tubes, each open at the downstream end, serviced, at their upstream ends by a single rotary valve, the axis of which lies parallel to the axes of the reactor tubes. The valve controls the admission of fuel and oxidant, sequentially, to each reactor tube. In addition, the valve contains pockets charged with high temperature-and-pressure products of combustion, tapped from a previously fired tube, to serve as a precisely timed high-energy

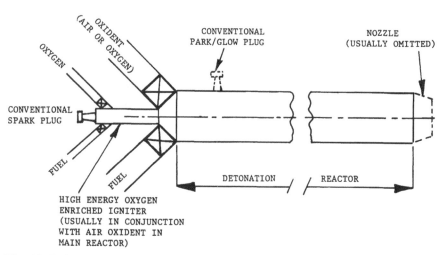

**Fig. 10–8** A generic, cyclic type, detonation tube reactor (diagrammatic).

igniter. The Edwards et al. device was intended to be suitable for a number of applications including use as a pressure-gain combustor for gas turbines.

Fairly recently a prototype single chamber detonation-type reactor has been operated experimentally by Helman et al. (1986). The Helman et al. reactor featured a very short reactor tube used in conjunction with an oxygen-enriched high-energy igniter. The fuel for both the igniter and the main detonation reactor was ethylene. The oxidizer in the main detonation reactor portion of the device was air. The maximum frequency of operation was only 25 Hz but this was thought to be due to the valve arrangement and it is expected that operation up to about 150 Hz should be feasible with a suitable valving system. More recent work by Eidelman et al. (1989, 1990) has extended the predictive aspects of the work, previously referred to, of Helman et al. The work of Eidelman et al. appears to confirm the prospects for a much higher operational frequency than the 25 Hz of the Helman et al. experiments. The main configurational difference between the Helman et al. and Eidelman et al. concepts was the provision, for the latter, of an annular, permanently open, inlet port at the head end of the reactor tube. The Eidelman et al. units were conceived, essentially, as potential nonresonant pulse jets.

The advantages claimed for detonation-type nonresonant pulse jets are relatively low specific fuel consumptions compared with those of competing devices and also for very short, high-frequency, detonation pulse jets, great compactness. Compared with, say, conventional aerovalved pulse jets the detonative units appear to promise substantially reduced specific fuel consumptions but at the price of the complications associated with valving and sequenced ignition. They also tend to be more compact than a pulse jet primarily because the long tailpipe of the latter is not required. However, for a given average thrust the maximum amplitude of the instantaneous thrust of a single-chamber detonative engine exceeds that of a single-chamber pulse jet by at least one order of magnitude. Both engines can be expected to operate, typically, in the 100 to 200 Hz frequency range.

A detonative gas-turbine pressure-gain combustor employing detonation waves that traverse laterally, rectangular combustion spaces has been suggested by Wortman (1984(a); 1984(b)). The essential features of the Wortman combustor are illustrated, diagrammatically, in Fig. 10.9. It can be seen that only a portion of the compressor outflow passes through the cluster of detonation combustors, the remainder passes through a conventional combustion system. The flows are reunited in the mixing region where the flow from the detonative combustors effectively pumps the remaining flow to establish a stagnation pressure level at entry to the turbine in excess of that prevailing at the compressor exit. The expansion zone downstream of the detonative reactors is intended to weaken the residual waves in the flow approaching the turbine. A very substantial area ratio contraction at the entry of each detonation tube is intended to protect the compressor from disturbances propagated upstream from the detonative reactors.

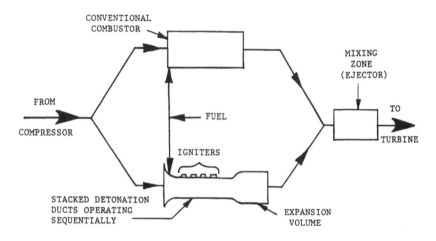

**Fig. 10–9** Diagrammatic illustration of a pressure-gain gas-turbine combustor employing transverse detonation waves. Proposed by Wortman, 1984.

### 10.3.3   Pressure-gain performance potential

A question that can be asked is what, in a general sense, is the performance potential for detonation-wave, pressure-gain combustors for gas turbines? At least an indication of the likely performance potential of this class of device can be obtained, without detailed analyses having been performed, by analogy with the expected performances of pulse combustor and also pressure-exchanger type pressure-gain combustors. Reasonable estimates of the performance of both the latter type of combustors have been established on the basis of experiments and fairly detailed theoretical predictions. Since detonation-wave combustion is, inherently, a close, transient, representation of constant volume combustion, and also since the combustion reaction can take place in unvitiated air, it might, perhaps, be expected that detonation-wave combustors would produce substantial pressure gains. However, consideration must be given to both the need to dampen the strong pressure peaks produced and also to entrain, and pump, the secondary air. Both of these processes involve significant irreversibilities.

By comparison a pulse, pressure-gain combustor is less effective as a constant volume system, largely because combustion takes place in the presence of high temperature products of combustion from previous cycles. The pulse combustor also needs to entrain, and pump, secondary air in the manner of a detonation-wave combustor. It would appear, therefore, that a well-designed detonation-wave combustor may have a slightly better performance than is achievable with a pulse, pressure-gain combustor. The expectation is, therefore, of a performance approximately comparable with that of a pressure-exchanger type pressure-gain combustor that, like the detonation wave device, carries out combustion in unvitiated air either at constant volume or at a condition simulating constant volume combustion (Section 7.4). A more detailed discussion of this topic is available (Kentfield and O'Blenes, 1988).

### 10.4   Integral Pressure-Exchanger Supercharger

Rarely, it seems, is it possible to incorporate a satisfactory pressure-exchanger as an integral part of another piece of equipment. A notable exception is the concept of an integral pressure-exchanger supercharger, originated by Curtil (1975), for use specifically on twin-cylinder two-stroke reciprocating engines. The Curtil exhaust-gas-energized supercharger consists, in the most basic form, of a tube interconnecting the two cylinders of an otherwise conventional, 180° firing angle, two-stroke at port height plus two pockets, one in each piston, each capable of connecting one end of the tube to an air inlet port. The arrangement is shown, diagrammatically, in Fig. 10.10. For an engine which is blower scavenged the incorporation of a Curtil supercharger does not introduce any additional moving parts.

   In operation the descending piston, referring to Fig. 10.10, uncovers the top edge of the supercharger port at, say, 75° after top-dead-center (TDC) on the expansion stroke. The opposite, ascending, piston, which is 180° out of phase, has just reached the bottom edge of the opposite supercharger port. The exhaust gases at high-pressure compress air in the supercharger tube, and push it into the cylinder in which the piston is ascending. Before exhaust gases can travel the length of the supercharger tube and enter the cylinder in which the piston

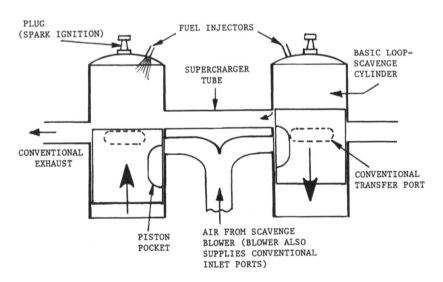

**Fig. 10–10** A Curtil supercharger tube applied to an otherwise conventional spark-ignition, or Diesel, twin-cylinder two-stroke engine (Curtil, 1975).

is ascending the piston in that cylinder cuts off further communication with the supercharger tube. Meanwhile the descending piston descends yet further, relieving the pressure in the supercharger tube via the conventional exhaust port. Continuing ascent of the opposite piston brings that piston pocket into a position permitting scavenge air to flow from left to right through the supercharger tube pushing exhaust products, at low pressure, into the right-hand cylinder and thus out through the conventional exhaust port of that cylinder. The whole sequence of events is, of course, reversed when the left-hand piston descends on the expansion stroke and eventually uncovers the top edge of the supercharger port.

It is worth noting that the Curtil device has the disadvantage, relative to a conventional pressure-exchanger, of relatively prolonged supercharger-tube opening and closing events. However, the comparatively high exhaust gas pressure prevailing at port opening, not available to a conventional pressure-exchanger supercharger or turbocharger, should compensate for this. The height of the supercharging ports corresponds, for an opening or closing 75° after or before TDC, to 30° of crankshaft rotation. For an opening and closing 80° after and before TDC, respectively, the port height corresponds to 20° of crankshaft rotation. It should be possible to minimize the intensity of mixing of exhaust gas and air within the supercharger tube due to slow port opening and closing by means of partitioning surfaces. The length of a Curtil supercharger tube depends upon the engine speed at which optimum operation is required. It was found, from a method-of-characteristics wave diagram study, that a tube of about 0.3 m (1 ft) length was suitable for an engine operating at 3000 rev/min. Clearly the system has the potential for application to engines with any even number of cylinders made up of pairs of alternatively firing cylinders.

# REFERENCES

Azoury, P. H., 1966, "An Introduction to the Dynamic Pressure Exchanger," *Proceedings*, Institution of Mechanical Engineers, Vol. 180, Pt. 1, No. 18, pp. 451–480.

Bannister, F. K., 1964, "Influence of Pipe Friction and Heat Transfer on Pressure Waves in Gases: Effects in a Shock Tube," *Journal of Mechanical Engineering Science*, Institution of Mechanical Engineers, Vol. 6, No. 3, pp. 278–292.

Benson, R. S., 1975, Research Note: "A Simple Algorithm for a Multi-Pipe Junction in Non-Steady Homentropic Flow," *Journal of Mechanical Engineering Science*, Institution of Mechanical Engineers, Vol. 17, No. 1, pp. 40–41.

Benson, R. S., Garg, R. D., and Woollat, D., 1964, "A Numerical Solution of Unsteady Flow Problems," *International Journal of Mechanical Engineering Science*, Pergamon Press, Vol. 6, No. 1, pp. 117–144.

Benson, R. S., Woods, W. A., and Woollatt, D., 1964, "Unsteady Flow in Simple Branch Systems," *Proceedings*, Institution of Mechanical Engineers, Vol. 178, Pt. 3, Paper 10, pp. 24–49.

Birmann, R., 1956, "Aerothermodynamic Considerations Involved in Turbocharging Four and Two-Cycle Diesel Engines," *Transactions*, ASME, Vol. 78, No. 1, pp. 171–183.

Brenchley, D. L., and Bomelburg, H. J., 1984, Pulse Combustion—An Assessment of Opportunities for Increased Efficiency. Report PNL-5301, UC-95, Battelle Pacific Northwest Laboratory.

Briffa, F. E. J., and Romaine, D. R., 1972, "Experiments with Coupled Pulse Combustors," *Proceedings*, Second Conference of Natural Gas Research Technology, Atlanta, Georgia, Paper No. 2, Session 1, American Gas Association.

Clarke, P. H., and Craigen, J. G., 1976, "Mathematical Model of a Pulsating Combustor," Paper C54/79, Sixth Thermodynamics and Fluid Mechanics Convention, Institution of Mechanical Engineers.

Corliss, J. M., and Putnam, A. A., 1985, Basic Research on Pulse Combustion Phenomena, Annual Report, Nov. 1983–Nov. 1984, GRI 85/0029, Apr. Gas Research Institute.

Cotter, J., 1963, private communication.

Courant, R., and Friedrichs, K. O., 1948, *Supersonic Flow and Shock Waves*, Interscience Publishers, New York.

Croes, N., 1979, "Die Wirkungsweise der Taschen des Druckwellenladers Comprex," *MTZ*, 40 DK 621.43.052.

Cronje, J. S., 1979, "An Experimental and Theoretical Study, Including Frictional and Heat Transfer Effects, of Pulsed Pressure-Gain Combustion," Ph.D. thesis, Department of Mechanical Engineering, University of Calgary, Calgary, Alberta, Canada.

270

Cronje, J. S., and Kentfield, J. A. C., 1980, "A Numerical Procedure for the Analysis of One-Space-Dimensional Non-Steady Compressible Flow," *Proceedings*, 7th Australasian Hydraulics and Fluid Mechanics Conference, Brisbane. Institution of Engineers of Australia publication No. 80/4, Aug., pp. 346–349.

Curtil, R., 1975, private communication.

Edelman, L. B., 1947, "The Pulsating Jet Engine—its Evolution and Future Prospects," *SAE Quart. Trans.*, Vol. 1, pp. 204–216.

Edwards, B. D., Lang, T. J., and Hill, R. J., 1969, Detonation Apparatus. British Patent Specification No. 1 269 123, filed 29 October.

Edwards, B. D., 1976, "Maintained Detonation Waves in an Annular Channel: A Hypothesis Which Provides the Link Between Classical Acoustic Combustion Instability and Detonation Waves," *Proceedings*, 16th International Symposium on Combustion, The Combustion Institute, pp. 1611–1618.

Eidelman, S., Grossman, W., and Lottati, I., 1989, "A Review of Propulsion Applications of the Pulsed Detonation Engine Concept," AIAA Paper No. AIAA-89-2446.

Eidelman, S., and Grossman, W., 1990, "Computational Analysis of Pulsed Detonation Engines and Applications," AIAA Paper No. AIAA-90-0460.

Foa, J. V., 1955, New Methods of Energy Exchange Between Flows and Some of its Applications. Technical Report TR AE 5509, Dec., Rensselaer Polytechnic Institute.

Foa, J. V., 1960, *The Elements of Flight Propulsion*, Wiley, London and New York.

Foa, J. V., 1970, "A Pressure Exchanger for Marine Propulsion," SAE Paper No. 700095.

Gibson, A. H., 1954, *Hydraulics and its Applications*, Constable and Co., London.

Glass, I. I., 1958, "Shock Tubes, Part 1. Theory and Performance of Simple Shock Tubes," *UTIA Review*, No. 12, Part 1. Institute of Aerophysics, University of Toronto, Toronto, Ontario, Canada.

Hanby, V. I., 1969, "Convective Heat Transfer in a Gas-Fired Pulsating Combustor," ASME *Journal of Engineering for Power*, Vol. 91, Series A, pp. 48–52; also, ASME Paper 68-WA/FU-1, 1968.

Hartree, D. R., 1958, *Numerical Analysis*, 2nd Edition, Oxford University Press.

Helman, D., Shreeve, R. P., and Eidelman, S., 1986, "Detonation Pulse Engine," AIAA Paper No. AIAA-86-1683.

Jenny, E., 1950, "Unidimensional Transient Flow with Consideration of Friction, Heat Transfer, and Change of Section," *Brown Boveri Review*, Vol. 37, No. 11, Nov., pp. 447–461.

Jonsson, V. K., Matthews, L., and Spalding, D. B., 1973, "Numerical Solution Procedure for Calculating the Unsteady Flow of Compressible Fluid (With Allowance for the Effects of Heat Transfer and Friction)," ASME Paper No. 73-FE-30.

Karnopp, D. C., Dwyer, H. A., and Margolis, D. L., 1975, "Computer Predictions of Power and Noise for Two-Stroke Engines with Power Tuned, Silenced, Exhausts," SAE Paper No. 750708.

Kearton, W. J., and Keh, T. H., 1952, "Leakage of Air Through Labyrinth Glands of Staggered Type," *Proceedings*, Institution of Mechanical Engineers, Vol. 166, p. 180.

Kentfield, J. A. C., 1963, "An Examination of the Performance of Pressure-Exchanger Equalisers and Dividers," Ph.D. thesis, Dept. of Mechanical Engineering, Imperial College, University of London.

Kentfield, J. A. C., 1968, "An Approximate Method for Predicting the Performance of Pressure Exchangers," ASME Paper No. 68-WA/FE-37.

Kentfield, J. A. C., 1969, "The Performance of Pressure-Exchanger Dividers and Equalisers," ASME *Journal of Basic Engineering*, Vol. 91, Series D, No. 3, pp. 361–368.

Kentfield, J. A. C., and Barnes, J. A., 1976, "The Pressure Divider: A Device for Reducing Gas-Pipe Line Pumping-Energy Requirements," *Proceedings*, 11th Intersociety Energy Conversion Engineering Conference, AIChE, pp. 636–643.

Kentfield, J. A. C., 1977, "A New Light-Weight Warm-Air Blower for Rapid Defrosting of Cold-Soaked Equipment," ASME Paper No. 77-WA/HT-20.

Kentfield, J. A. C., and O'Blenes, M., 1988 "Methods for Achieving a Combustion-Driven Pressure Gain in Gas in Gas Turbines," ASME *Journal of Engineering for Gas Turbines and Power*, Vol. 110, Oct., pp. 704–711.

Kentfield, J. A. C., and Fernandes, L. C. V., 1990(a), "Improvements to the Performance of a Prototype Pulse, Pressure-Gain, Gas Turbine Combustor," ASME *Journal of Engineering for Gas Turbines and Power*, Vol. 112. Jan., pp. 67–72; also, ASME Paper No. 89-GT-277.

Kentfield, J. A. C., and Fernandes, L. C. V., 1990(b), "Further Development of an Improved Pulse, Pressure Gain, Gas Turbine Combustor," ASME Paper No. 90-GT-84.

Kirsten, W., 1971, "Influence of Exhaust Manifold Arrangements on Blade Vibrations in Turbocharger Turbines," *Brown Boveri Review*, Vol. 58, No. 4/5, Apr./May, pp. 145–160.

Lax, P. D., and Wendroff, B., 1960, "Systems of Conservation Laws," *Communications on Pure and Applied Mathematics*, Vol. 13, pp. 217–237.

Lockwood, R. M., Sargent, E. R., and Beckett, J. E., 1960, Thrust Augmented Intermittent Jet Lift-Propulsion System "Pulse Reactor." Report No. ARD-256 (final report), Hiller Aircraft Corporation.

Lord Rayleigh, 1945, *The Theory of Sound, Vol. 2*, Reprinted by Dover, New York.

Magnet, J. L., and Curtil, R., 1980, "Modular Pulse Converter Turbocharging Process and Exhaust Valve Facing Temperature," ASME Paper No. 80-DGP-4.

Marchal, R., and Servanty, P., 1963, "Note on the Development of Pulse Jets Without Flap Valves [Note sur le Développement de Pulso-Réacteurs Sans Clapets]," *Proceedings of Association Technique et Maritime et Aeronautique*, pp. 1–20 (English translation by Council for Scientific and Industrial Research, P.O. Box 395, Pretoria, South Africa, 1970).

Martin, J. F., 1960, "New Tool for Research," *Research Trends*, Vol. 7, No. 4. Cornell Aeronautical Laboratory Inc.

Matthews, L., 1969, "An Algorithm for Unsteady Compressible One-Dimensional Flow," Technical Note UF/TN/E/3, Oct., Dept. of Mechanical Engineering, Imperial College, University of London.

Meier, E., 1968, "Applications of Pulse Converters to Four-Stroke Diesel Engines with Exhaust-Gas Turbocharging," *Brown Boveri Review*, Vol. 55, No. 8, Aug., pp. 414–419.

Meier, E., 1971, "New Exhaust Systems for Turbocharged Internal Combustion Engines," *Brown Boveri Review*, Vol. 58, No. 4/5, Apr./May, pp. 148–160.

Nicholls, J. A., Wilkinson, H. R., and Morrison, R., 1957, "Intermittent Detonation as a Thrust Producing Mechanism," *Jet Propulsion*, Vol. 27, May, pp. 534–541.

Olorunmaiye, J. A., and Kentfield, J. A. C., 1989, "Numerical Simulation of Valveless Pulsed Combustors," *Acta Astronautics*, Vol. 19, No. 8, pp. 669–679.

Persechino, M. A., 1957, "Valveless Pulsejet Deicer Application," Report No. 5024, Nov., U.S. Naval Research Laboratory.

Petak, H., 1964, "Erfahrungen mit einfachen Pulse-Convertern an Viertakt-Dieselmotoren," *MTZ*, Vol. 25, No. 5, pp. 202–8; also, Sulzer Technical Review, Research Number, 1966, pp. 49–57.

Porges, G., 1977, *Applied Acoustics*, Halstead Press, Wiley, New York.

Porter, C. D., 1958, "Valveless-Gas-Turbine Combustors With Pressure Gain," ASME Paper No. 58-GTF-11.

Putnam, A. A., 1971, *Combustion Driven Oscillations in Industry*, American Elsevier.

Putnam, A. A., Belles, F. E., and Kentfield, J. A. C., 1986, "Pulse Combustion," *Progress in Energy and Combustion Science*, Vol. 12, pp. 43–79.

Reader, G. T., 1977, "The Pulse Jet, 1906–1966," *Journal of Naval Science*, Vol. 3, No. 4, pp. 226–232.

Riemann, G. B., 1858/59, "Über die Fortpflanzung ebener Luftwellen von endlicher Schwingungsweite," *Gött. Abh.*, Vol. 8 (Math.), pp. 43–65.

Rijke, P. L., 1859, "Nature of New Method of Causing a Vibration of the Air Contained in a Tube Open at Both Ends," *Phil. Mag.*, Vol. 17, p. 419.

Riley, K. F., 1974, *Mathematical Methods for the Physical Sciences; an Informal Treatment for Students of Physics and Engineering*, Cambridge University Press.

Rudinger, G., 1969, *Non-Steady Duct Flow: Wave Diagram Analysis*, Dover Publications, New York.

Ruf, W., 1967, "Berechnungen und Versuche an Druckwellen-Maschinen unter besonderen Berücksichtigung des Druckteilers und Injektors," Ph.D. thesis, Eidgenossischen Technischen Hochschule (E.T.H.), Zürich.

Servanty, P., 1968, "Réflexions sur la Combustion Pulsatoire," *Entropie*, No. 22, July–Aug., pp. 49–63.

Shapiro, A. H., 1953, *The Dynamics and Thermodynamics of Compressible Fluid Flow, Vol. 2*, Ronald Press, New York.

Schchelkin, K. I., and Troshin, Ya. K., 1965, *Gasdynamics of Combustion*, Mono Book Corporation (English translation by Kuvshinoff, B. W. and Holtschlag, L.)

Smith, W. E., and Weatherston, R. C., 1958, Studies of a Prototype Wave Superheater. Report No. HF-1056-A-1, AFOSR TR 58-158 AD 207244, Dec. Cornell Aeronautical Laboratory (now Calspan Advanced Technology Center, Buffalo, N.Y.).

Sneddon, I. N., 1957, *The Elements of Partial Differential Equations*, McGraw-Hill, New York.

Spalding, D. B., 1956, Theoretical Consideration of Leakage Effects in Pressure Exchangers. Report No. 2224/X47, Oct., Power Jets (R and D) Ltd.

Spalding, D. B., 1969, "A Procedure for Calculating Unsteady, One-Dimensional, Flow of a Compressible Fluid with Allowance for the Effects of Heat Transfer and Friction," Technical note UF/TN/D/2, Nov., Dept. of Mechanical Engineering, Imperial College, University of London.

Speirs, B. C., 1989, "An Experimental Optimisation of a Large, Aerovalved, Multiple Inlet Pulse Combustor," M.Sc. dissertation, Dept. of Mechanical Engineering, University of Calgary, Calgary, Alberta, Canada.

Spiers, H. M., 1955, *Technical Data on Fuel*, British National Committee World Power Conference, London.

Sran, B. S., and Kentfield, J. A. C., 1982, "Twin Valveless Pulse Combustors Coupled to Operate in Antiphase," *Proceedings*, Vol. 1, Paper No. 3, *Symposium of Pulse-Combustion Applications*, Gas Research Institute/U.S. Dept. of Energy.

Streeter, V. L. and Wylie, E. B., 1981, *Fluid Mechanics*, McGraw-Hill Ryerson.

Summerauer, I., Spinnler, F., Mayer, A., and Hofner, A., 1978, "A Comparative Study of the Acceleration Performance of a Truck Diesel Engine with Exhaust-Gas Turbocharger and with Pressure-Wave Supercharger Comprex," Paper C70/78, Institution of Mechanical Engineers, London.

Thring, M. W., 1961, *The Collected Works of F. H. Reynst*, Pergamon Press, Ltd., London.

Warren, M. D., 1983, "Appropriate Boundary Conditions for the Solution of the Equations of Unsteady One-Dimensional Gas Flow by the Lax-Wendroff Method," *International Journal of Heat and Fluid Flow*, Butterworth and Co., Vol. 4, No. 1, Mar., pp. 53–59.

Weatherston, R. C., Smith, W. E., Russo, A. L., and Marone, P. V., 1959, Gas Dynamics of a Wave Superheater Facility for Hypersonic Research and Development. Report No. AD-1118-A-1, AFOSR TN 59-107, Feb. Cornell Aeronautical Laboratory (now Calspan Advanced Technology Center, Buffalo, N.Y.).

Wortman, A., 1984(a), Detonation Wave Augmentation of Gas Turbines. Report No. IST-NAS-07-84-01, Istar Inc., Santa Monica, CA 90402.

Wortman, A., 1984(b), "Detonation Wave Augmentation of Gas Turbines," AIAA Paper No. AIAA-84-1266.

Wylie, E. B., and Streeter, V. L., 1978, *Fluid Transients*, McGraw-Hill.

Zehnder, G., 1968, "Pulse Converters on Two-Stroke Diesel Engines," *Brown Boveri Review*, Vol. 55, No. 8, Aug., pp. 414–419.

Zehnder, G., Mayer, A., and Matthews, L., 1989, "The Free Running Comprex," SAE Paper No. 890452.

Zeller, H., 1965, "Strömungsvorgänge in Pulsierenden Arbeitenden Strahlapparaten," Abhandlungen aus dem Aerodynamischen Institut, T.H., Aachen, No. 18, pp. 20–25.

Zeller, H., 1966, "Eindimensionale Strömung in Strahlapparaten," Forschungsbericht des Landes Nordrhein-Westfalen No. 1658, Westdeutcher Verlag, Cologne.

Zhuber-Okrog, G., 1968, "Über die Wirkungsweise von Schmidt-Rohren." Forschungsbericht 68–49, Deutsche Luft und Raumfahrt.

Zhuber-Okrog, G., 1976, "Über die Vorgange in Strahlrohren mit Pulsierender Verbrennung," Fortschr.-Ber. VDI Z.6, No. 47.

Zinn, B. T., Miller, N., Carvalho, J. A., and Daniel, B. R., 1982, "Pulsating Combustion of Coal in a Rijke-Type Combustor," *Proceedings*, 19th International Symposium on Combustion, The Combustion Institute, pp. 1197–1203.

Fig. 5-10 Normalized boundary curves for emptying processes ($\gamma = 1.4$) (after Jenny).

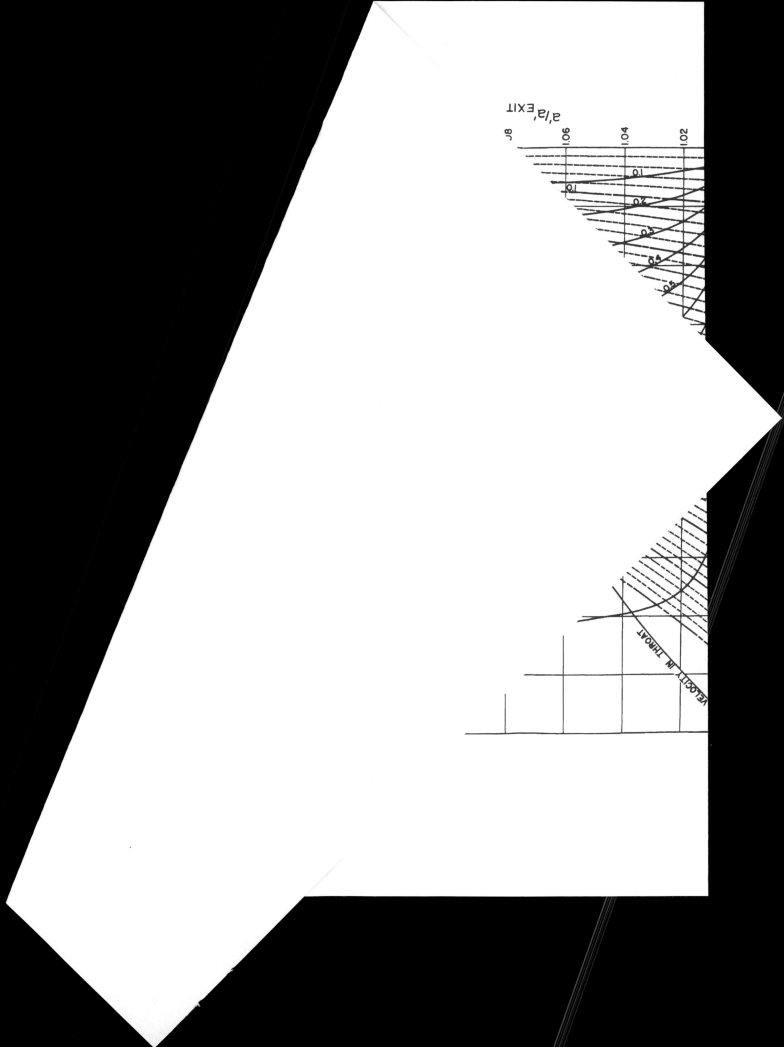

STEADY FLOW SUDDEN ENLARGEMENT BOUNDARY CURVES
FOR CELL FILLING

(TAKEN FROM LOWER R.H. PORTION OF JENNY'S FIG 23)

INLET
STAGNATION
CONDITIONS
{ $P_0'$ INLET
$a_0'$ INLET
($U_0'$ INLET $=0$) }

$\phi A$

$P'$
$a'$

$\longrightarrow U'$

A

NOTE DIMENSIONLESS MASS
FLOW, $\dot{m}_0'$ IN , IN CELL END
(& REFERRED TO INLET
STAGNATION CONDITIONS)
IS DEFINED THUS:

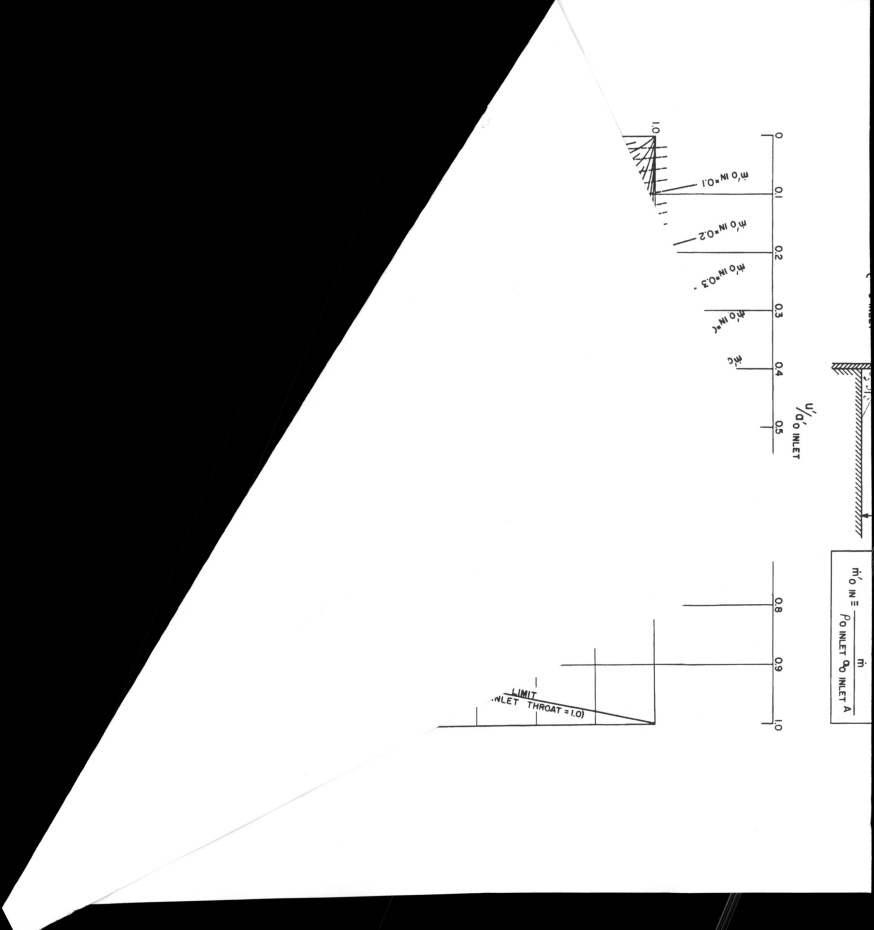